Mental Exercise for Dog

Crafting the Perfect Play Routine. Techniques and games Top Trainers Use to Ensure a Balanced, Happy, and Mentally Stimulated Dog

Amelia Steam

Table of Content

FOREWORD...9

THE EVOLUTION OF CANINE PLAY AND MENTAL STIMULATION..9

INTRODUCTION..13

THE INTEGRAL ROLE OF MENTAL STIMULATION IN A DOG'S LIFE...13

THE SCIENCE OF CANINE COGNITION AND PLAY...14

CHAPTER 1: THE CANINE BRAIN UNVEILED...19

ANATOMY OF THE CANINE BRAIN...19

The Cerebrum: The Center of Thought and Action 20

The Cerebellum: Maintaining Balance and Grace 20

The Brainstem: The Lifeline...20

COGNITIVE DEVELOPMENT ACROSS DIFFERENT LIFE STAGES...21

Puppyhood: A World of Discovery 22

Adolescence to Adulthood: Refinement and Mastery 22

Senior Years: The Grace of Wisdom 23

BREED-SPECIFIC COGNITIVE TRAITS AND TENDENCIES..24

The Border Collie: A Symphony of Focus and Intelligence 24

The Bloodhound: Scent Wizards with a Memory Map 25

The Golden Retriever: Empathetic Companions 26

The Dachshund: Fierce and Independent Thinkers 27

CHAPTER 2: RECOGNIZING THE SIGNS..31

SYMPTOMS OF A MENTALLY UNDER-STIMULATED DOG 31

THE CONSEQUENCES OF NEGLECTING MENTAL EXERCISE 32

BEHAVIORAL ISSUES STEMMING FROM LACK OF MENTAL ENGAGEMENT 34

CHAPTER 3: LAYING THE FOUNDATIONS FOR EFFECTIVE PLAY 39

THE PSYCHOLOGY OF POSITIVE REINFORCEMENT 39

ESSENTIAL COMMANDS: BEYOND THE BASICS 41

BUILDING TRUST: THE CORNERSTONE OF INTERACTIVE PLAY...43

Fostering a Safe Environment 44

Consistency is Key 44

Listening with an Open Heart 44

Celebrate the Small Moments 45

Respecting Boundaries 45

CHAPTER 4: INTERACTIVE BRAIN GAMES 47

HIDE AND SEEK VARIATIONS: REDISCOVERING AN AGE-OLD GAME 47

The Classic Version with a Twist 47

Hide and Seek in the Dark 48

The Multiple Hiders Game 48

The Outdoor Challenge 48

Hide, Seek, and Command 48

Making it Rewarding 48

Hide and Seek: More than Just a Game 49

ADVANCED FETCH GAMES: BEYOND THE BASIC THROW 49

The World of Object Diversity: Size and Shape 49

Throwing with Panache: The Craft and Care 50

Turning Fetch into an Obstacle Adventure 50

The Adventure of Different Terrains 50

The Dance of Commands 51

Mysteries of the Night-time Fetch 51

Perfecting the Return and Release 51

PUZZLE TOYS: FROM STORE-BOUGHT TO DIY 51

A Glimpse into the World of Store-Bought Puzzles 52

Crafting the Challenge: The DIY Route 52

The Impact: Beyond Just Play 53

CHAPTER 5: SCENT WORK AND NOSE GAMES 55

THE SCIENCE OF CANINE OLFACTION 55

An Evolutionary Perspective 55

The Emotional Connection to Scents 56

Scent Memory: A Dog's Personal Diary 56

Decoding Pheromones: Silent Conversations 57

Environmental Influence on Olfaction 57

SETTING UP SCENT TRAILS AND TREASURE HUNTS 58

Crafting The Perfect Scent Trail 58

The Lure of the Treasure Hunt 58

Building on Instinct, Crafting Connection 59

Challenges and Safety Tips 59

A New Chapter in Play 59

ADVANCED NOSE WORK: CHALLENGES AND COMPETITIONS...60

   A Higher Calling: The Essence of Advanced Nose Work 60

   Crafting Challenges: From Training to Execution 61

   The Competitive Arena: Stepping into the Limelight 61

   The Bond Beyond the Game 62

   Navigating the Challenges: Ethical Considerations and Safety 62

   A Symphony of Scents: The Closing Note 63

CHAPTER 6: AGILITY AND PHYSICAL BRAIN BOOSTERS...65

   DESIGNING A HOME AGILITY COURSE...65

   INCORPORATING MENTAL CHALLENGES INTO PHYSICAL PLAY 67

   ADAPTING AGILITY FOR DIFFERENT DOG SIZES AND BREEDS 69

CHAPTER 7: SOCIAL PLAY AND GROUP CHALLENGES...75

   THE DYNAMICS OF DOG-TO-DOG PLAY 75

   ORGANIZING GROUP PLAY SESSIONS AND CHALLENGES 77

   ADDRESSING SOCIAL PLAY ISSUES: FROM SHYNESS TO AGGRESSION...79

   The Silent Observer: Addressing Shyness 80

   The Aggressive Playmate: Navigating the Storm 80

CHAPTER 8: ADVANCED TRAINING AND COGNITIVE CHALLENGES 83

   CLICKER TRAINING AS A BRAIN GAME...83

   TEACHING COMPLEX TRICKS AND SEQUENCES...85

   COGNITIVE TOYS AND TECH FOR DOGS 87

CHAPTER 9: CRAFTING TAILORED PLAY ROUTINES 91

   ASSESSING AND DOCUMENTING YOUR DOG'S PROGRESS 91

   ADAPTING PLAY FOR PUPPIES, ADULTS, AND SENIOR DOGS...92

   The Energetic Pup: Harnessing Boundless Energy 93

   The Dynamic Adult: Perfecting the Balance 93

   The Graceful Senior: Play with a Touch of Tenderness 93

   ADDRESSING SPECIFIC BEHAVIORAL CHALLENGES THROUGH PLAY 94

   The Tale of the Overly Energetic Mutt 94

   Mending the Bonds of Trust 95

   The Case of the Territorial Terrier 95

CHAPTER 10: OVERCOMING OBSTACLES IN MENTAL STIMULATION 99

   DEALING WITH RELUCTANT OR FEARFUL DOGS...99

   MODIFYING GAMES FOR DOGS WITH DISABILITIES...100

ENSURING SAFETY AND WELL-BEING DURING PLAY 102

CHAPTER 11: BEYOND THE HOME: EXTERNAL ADVENTURES AND CHALLENGES 107

ENRICHING WALKS AND OUTDOOR ADVENTURES 107

DOG SPORTS AND COMPETITIONS 108

TRAVEL AND EXPLORATION: MENTAL STIMULATION ON THE GO 111

CONCLUSION 115

THE LIFELONG COMMITMENT TO YOUR DOG'S MENTAL WELL-BEING 115

THE RIPPLE EFFECTS OF A MENTALLY STIMULATED DOG 117

APPENDIX A: COMPREHENSIVE TOY AND EQUIPMENT GUIDE 121

REVIEWS, RECOMMENDATIONS, AND SAFETY TIPS 121

Durability vs. Novelty 121

Interactive Toys – A Dual Purpose 121

Safety First 122

PERSONALIZED CHOICES 122

Recommendations and Endorsements 122

DIY PROJECTS AND IDEAS FOR HOME-MADE BRAIN GAMES 123

Crafting with Care 123

The Essence of Materials 123

Ingenious DIY Brain Games 124

The Beauty of Customization 124

Cherishing the Moments 125

In Retrospect 125

APPENDIX B: EXPERT INSIGHTS AND FURTHER LEARNING 127

INTERVIEWS WITH CANINE COGNITIVE SCIENTISTS AND TRAINERS 127

A Conversation with Dr. Sophie Tremblay: Canine Cognitive Scientist 127

Delving Deeper with Marco Stevenson: Veteran Dog Trainer 127

Insights from Clara Hughes: Specialist in Dog Behavior and Therapy 128

Final Reflections 129

RECOMMENDED READING, COURSES, AND ONLINE RESOURCES 129

A Dive into Literature: Essential Books for Every Dog Owner 129

Beyond Books: Engaging Courses to Elevate Your Knowledge 130

Digital Domains: Trustworthy Online Resources 130

APPENDIX C: CASE STUDIES 133

REAL-LIFE STORIES OF TRANSFORMATION THROUGH MENTAL PLAY 133

Lucy and Whiskey: A Second Chance at Connection 133

Jasper's Journey with Ethan: Unlocking Hidden Potentials 133

Rebecca and Daisy: From Fear to Confidence 134

Milo's Metamorphosis with Alex: From Chaos to Calm 134

ADDRESSING AND OVERCOMING SPECIFIC BEHAVIORAL ISSUES THROUGH MENTAL PLAY 135

The Tale of Bella: Conquering Separation Anxiety 135

Max's Story: Overcoming Obsessive Behaviors 136

Juno's Journey: Curbing Excessive Barking 136

Riley and the Battle with Aggression 136

Nina's Nights: The Struggle with Night-time Restlessness 137

Concluding Insights 137

Foreword

The Evolution of Canine Play and Mental Stimulation

The relationship between humans and dogs has evolved dramatically over millennia. From hunters and protectors to companions and family members, dogs have played diverse roles in our lives. But in recent years, a profound shift has occurred in how we approach our interactions with these faithful companions. This shift centers on recognizing and nurturing the intellectual and emotional facets of our dogs through play and mental stimulation.

Dogs are not just instinctual creatures responding to stimuli; they are beings with complex mental lives that thrive on challenge, curiosity, and engagement. They seek more than a game of fetch or a walk in the park. Modern dogs, just like their human counterparts, desire intellectual stimulation that resonates with their innate curiosity and need to explore.

Understanding canine intelligence is not just a trend but a movement towards recognizing our furry friends as sentient beings with desires and capacities for intellectual growth. The process of unraveling the mysteries of the canine mind takes us far beyond obedience and tricks, diving into empathy, shared joy, and the bond that comes from working through challenges together.

With advancements in animal psychology and behavior studies, we have begun to realize the immense cognitive potential within each dog. This potential is not uniform; it's unique to each individual, much like human intelligence. Whether it's an athletic retriever eager for mental and physical workout or a contemplative spaniel who finds joy in puzzles, recognizing these traits allows for an enriching and personalized connection.

The evolution of canine play and mental stimulation is about honoring the diversity of the canine world. It's about tailoring our interactions to meet the needs, capabilities, and desires of each unique pet. The techniques and philosophies presented in this book are more than methods; they are pathways to a deeper, more satisfying relationship with our four-legged friends.

Your journey into the captivating world of canine mental stimulation and play is about to begin. Whether you're a seasoned dog owner or just starting out, may this book serve as your guide, helping you to forge a connection with your dog that transcends the ordinary.

May the pages that follow enrich your understanding, deepen your bond, and lead you both to new horizons of joy, trust, and mutual growth.

Warmly,
Amelia Steam

Introduction

In our quest to understand the creatures that share our lives, few subjects are as enchanting and vital as the mental life of our beloved dogs. Science has taken giant leaps in recent years, giving us an unprecedented window into the canine mind. The mysteries of play, once seen as mere amusement, now unfold as a rich tapestry of cognition, emotion, and connection. This book unravels the science of canine cognition and play, taking you on a journey that promises to transform the way you see and interact with your furry friend.

The Integral Role of Mental Stimulation in a Dog's Life

Our history with dogs weaves a tapestry rich in companionship and mutual growth. Thousands of years of evolution have fine-tuned this bond, taking it beyond mere domestication. Today, it's a relationship marked by understanding, empathy, and love. One of the most profound realizations in our ongoing journey with these loyal creatures is the importance of mental stimulation for their holistic well-being.

It's easy to fall into the trap of seeing dogs as creatures of habit and instinct alone, especially when their daily lives often revolve around routine. However, this is a simplistic view. The truth is, dogs are deeply curious and intelligent beings. Their minds, active and engaged, brim with emotions, thoughts, and desires that need nurturing. And just like us, they seek purpose, challenges, and opportunities to learn.

Play, often mistaken as a mere pastime, holds profound significance in the life of a dog. Beyond the physical release it offers, play represents a cerebral journey. When a dog engages in play, it's not just about chasing a ball or tugging at a toy; it's a venture into problem-solving, social interactions, and creative thinking. It's an exploration of their environment, tapping into their innate creativity and adaptability.

When we teach our dogs new commands or tricks, we're doing more than training them to follow orders. This learning process is a gateway to cognitive development. It enhances their adaptability, lays the foundation for problem-solving skills, and most importantly, it strengthens the emotional bridge between us and our furry friends.

Incorporating puzzles or challenges into a dog's routine can be a game-changer. Imagine the joy and sense of accomplishment they feel when they successfully navigate a new challenge. This not only sharpens their minds but also gives them a sense of purpose and achievement.

Another critical aspect of a dog's mental stimulation is their social interactions. Dogs are pack animals by nature. Their interactions with other dogs, animals, and, most importantly, humans, significantly influence their emotional health. Engaging in cooperative activities or simply spending quality time together can foster empathy, cooperation, and emotional understanding.

Moreover, the significance of mental stimulation doesn't wane as a dog ages. While the needs and capacities might shift over time, from the eager learning phase of a puppy to the mature contemplations of an older dog, the desire for mental engagement remains constant. Providing consistent mental stimulation can also have beneficial effects on a dog's overall health, reducing anxiety, curbing destructive behaviors, and fostering a sense of contentment.

In essence, understanding the intricate world of canine cognition and the role of mental stimulation reshapes our perspective. Taking care of a dog is not a linear task confined to feeding and physical exercise. It's a multifaceted commitment that involves nurturing their minds, recognizing their cognitive needs, and ensuring they lead fulfilling, enriched lives.

Our shared journey with dogs is more than a tale of master and pet; it's a partnership, a mutual exploration of life's complexities. And as we turn the pages of this book, we'll delve deeper into understanding these complexities and learning how to foster a nurturing environment for our beloved canine companions.

The Science of Canine Cognition and Play

When you gaze into the eyes of a dog, it's impossible to deny the depth and emotion that emanates from those soulful orbs. Anyone who's spent considerable time with a canine companion knows the sharpness of their minds and the weight of their emotions. This understanding, however, is no longer just an anecdotal consensus shared among pet owners. It's backed by robust scientific research.

Let's journey into the world of canine cognition, a realm where researchers and dog lovers unite to decode the mysteries that lie beneath those furry brows. While the concept might seem novel to some, in truth, the scientific community has long been intrigued by the inner workings of the canine mind.

Dogs, domesticated from their wild ancestors over thousands of years, have developed cognitive faculties remarkably attuned to human social cues. These capabilities aren't merely the byproduct of domestication; they are a testament to the profound relationship between our species.

A fascinating study that sheds light on their cognitive prowess is the 'pointing experiment'. When a human points to an object, even a puppy, with minimal exposure to humans, will follow the gesture. This might seem simple, but it's an ability even our closest primate relatives struggle with. Such innate understanding of human gestures suggests that dogs have evolved specialized skills to read and interact with us.

Delving deeper into their cognitive capabilities, researchers have uncovered dogs' potential for episodic memory. This means that dogs don't just live in the 'here and now'; they can recall past events, even if they weren't particularly significant at the time. Such findings challenge the long-standing belief that non-human animals live solely in the present, reaffirming the complexity of the canine mind.

The realm of play, often considered a leisurely activity, holds paramount importance in understanding canine cognition. To us, a game of fetch or tug-of-war might seem like a simple diversion, but to researchers, these interactions are a goldmine of information. Through play, dogs showcase problem-solving skills, creativity, and even their capacity for fairness. For instance, dogs have a sense of 'turn-taking' during play, an ability once thought unique to primates.

Understanding canine play isn't just about decoding their behaviors but also about appreciating the neurochemical changes occurring beneath the surface. Dopamine, the 'feel-good' neurotransmitter, sees a significant surge during play, affirming its role not just in physical activity but in mental well-being.

Another incredible facet of canine cognition is their emotional intelligence. Dogs can read human emotions, respond to our feelings, and even exhibit empathy. They comfort us when we're down, share in our joy, and, at times, even mirror our emotional states. Such deep-seated emotional connections are not merely products of training; they are ingrained in the very fabric of their being.

In recent years, advanced imaging techniques, such as fMRI scans, have allowed us a glimpse into the active canine brain. These studies, while in their nascent stages, have already revealed that dogs process information in ways more complex than previously imagined. The regions of their brains that light up in response to human voices, praise, or the scent of a familiar person are remarkably similar to our own, further cementing the notion that our bond is built on mutual understanding and shared emotions.

As we delve deeper into the book, we'll uncover how this understanding of canine cognition directly influences their need for mental stimulation. We'll learn how to harness their cognitive and emotional strengths to foster a more profound, more enriching bond. Because, when we truly understand the science behind their thoughts, emotions, and behaviors, we're better equipped to cater to their needs and ensure their well-being.

The exploration of canine cognition and play opens up a new perspective on our understanding of dogs. They are not mere animals responding to stimuli, but intelligent creatures capable of complex emotions, learning, and communication. Engaging with them through play is not only a source of entertainment but a powerful tool for connection and growth. By embracing this new perspective, we empower ourselves to forge deeper, more meaningful relationships with our canine companions, enriching our lives and theirs. The following chapters will provide practical guidance, innovative techniques, and real-world insights to bring this exciting science to life.

As we conclude this introduction, we're reminded of the profound and intricate dance we share with our canine companions. A dance not of steps, but of heartbeats, emotions, and shared glances. Their world, rich in sensory experiences, is a testament to evolution, adaptation, and above all, our intertwined destinies. Our history with dogs spans thousands of years, but only recently have we truly begun to unlock the depths of their minds.

Each soulful gaze and joyful bark is not just a fleeting moment but a window into a world where cognition and emotion meld seamlessly. We've barely scratched the surface of understanding the sheer depth of their intellect and their incredible capacity to resonate with human emotions. And while science continues to unravel the mysteries of their minds, it's evident that our bond with dogs is built on mutual respect, trust, and a deep-seated desire for companionship.

As we venture further into this book, we're not just on a quest for knowledge; we're embarking on a journey of the heart. A journey that seeks to deepen our connection, enhance our shared experiences, and celebrate the joyous dance of human and canine minds intertwining. Every page turned, every story shared, and every lesson learned is a step closer to understanding the very essence of this unique bond. A bond that doesn't merely exist, but thrives in the shared moments, mutual discoveries, and the infinite love we share with our dogs.

In the chapters that lie ahead, we'll explore, discover, and learn together. More than just a guide, this book is an ode to the incredible relationship we share with our canine companions. So, as we take this leap into the intricate tapestry of canine cognition and emotion, let's remember: it's not about merely understanding our dogs but about forging an even deeper, more meaningful connection with them. Here's to the journey ahead and the shared discoveries that await us.

Chapter 1: The Canine Brain Unveiled

The wondrous realm of canine cognition is as vast as the stars in the night sky. Each twinkle, every glimmer represents a unique facet of a dog's mind, shaped by both nature and nurture. This chapter delves into the intricate design of the canine brain, uncovering the symphony of its anatomy, the stages of cognitive development, and the captivating dance of breed-specific traits. As we journey together, imagine yourself as a traveler, equipped with a compass of curiosity, ready to chart the profound depths of the canine psyche.

Anatomy of the Canine Brain

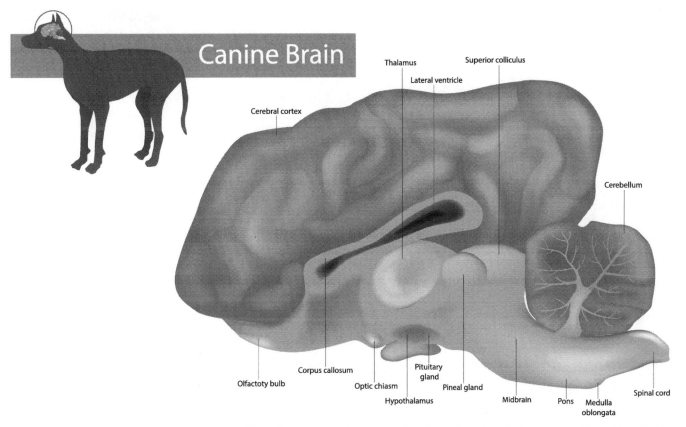

As the sun sets, casting a warm, golden hue over the neighborhood, a local dog named Rufus finds himself in his favorite spot at the window, intently watching the world go by. A squirrel dashes across the lawn, and Rufus's ears perk up. A child's laughter echoes from a nearby yard, and his tail wags in response. The aroma of a barbeque wafts in, making him drool just a bit. But have you ever paused to wonder how Rufus processes all this information? What goes on inside that keen mind of his? It all begins with the intricate anatomy of the canine brain.

Peering inside a dog's skull, one would find a brain that, at first glance, might not look vastly different from our own. But as with many things in life, the devil is in the details.

The canine brain, like the human counterpart, is divided into different regions, each responsible for specific functions. The most prominent of these regions are the cerebrum, cerebellum, and brainstem.

The Cerebrum: The Center of Thought and Action

When we think of a dog's personality, behaviors, and ability to learn new tricks, we're largely attributing these feats to the cerebrum. This part of the brain, with its intricate folds and crevices, is the largest region and is primarily responsible for voluntary muscle activities, interpreting sensory information, and higher-order functions like thinking and decision-making.

The outer layer of the cerebrum, known as the cerebral cortex, is a densely packed layer of nerve cells. This is where the magic truly happens. It's where Rufus identifies the scent of the barbeque or the sound of the laughing child. Within the cerebral cortex lies areas dedicated to specific tasks. There are regions for vision, hearing, smell, and so forth. It's like a well-orchestrated symphony, where each section plays its part to create a harmonious output.

The Cerebellum: Maintaining Balance and Grace

Just below the cerebrum, you'll find the cerebellum, a region that might be smaller in size but is monumental in function. If you've ever watched a dog gracefully chase a ball, take a sharp turn, or balance on their hind legs for a treat, you've witnessed the cerebellum in action. This part of the brain controls balance, coordination, and smooth muscle activity. For Rufus, it's what ensures he doesn't tumble over while fervently chasing that darting squirrel.

The Brainstem: The Lifeline

Connecting the brain to the spinal cord, the brainstem is like the control center for many of the body's vital functions. It's where critical processes like breathing, heart rate, and blood pressure are regulated. Though we might not always acknowledge it, the brainstem's functions are the unsung heroes of existence, continuously working in the background, ensuring everything runs smoothly.

Now, within these broad regions are more specific structures that play pivotal roles in a dog's life. The olfactory bulb, for instance, is a testament to a dog's unparalleled sense of smell. Humans, with our mere 6 million olfactory receptors, pale in comparison to the average dog, boasting around 300 million. This structure processes scent information, making sense of the myriad smells that dogs encounter daily.

Another remarkable structure is the amygdala. Tucked deep within the cerebrum, the amygdala plays a crucial role in processing emotions, especially those linked to survival like fear and aggression. When a dog feels threatened or anxious, it's the amygdala that triggers the appropriate response, be it a bark, growl, or retreat.

Diving even deeper, one would find intricate networks of neurotransmitters — chemicals responsible for transmitting signals between nerve cells. Serotonin, dopamine, and oxytocin are just a few of these. When Rufus feels the joy of playing fetch or the comfort of his owner's touch, these neurotransmitters surge, creating feelings of happiness and contentment.

In conclusion, understanding the anatomy of the canine brain isn't just a scientific endeavor. It's a journey into the essence of what makes our furry friends tick. It's realizing that beneath that playful exterior lies a complex, dynamic organ that feels, thinks, and even dreams. As we move forward, we'll delve deeper into how this anatomy plays out in different life stages and among various breeds, offering us an even more profound insight into the world of dogs.

Cognitive Development Across Different Life Stages

In the soft glow of the early morning sun, Lucy, a playful Golden Retriever puppy, stumbles around the garden. Her world is vast, filled with wonder. Each blade of grass is a new sensation under her tiny paws, every rustling leaf a source of amazement. Fast forward a few years, and Lucy, now in her prime, confidently navigates the same garden, responding promptly to commands, and displaying an intelligence that makes her family beam with pride. Eventually, as the years wear on, Lucy becomes a serene elder, her once rapid responses slightly dulled, but her wisdom evident in every gaze.

Just like humans, dogs traverse a compelling journey of cognitive development that spans from their puppyhood to their golden years. Each stage of their life brings about unique challenges and feats, painting a picture of growth, adaptation, and evolution.

Puppyhood: A World of Discovery

Imagine waking up each day to a world bursting with novel sights, sounds, and scents. That's the life of a puppy. This phase, stretching from birth to about six months, is marked by intense learning and exploration. The neural pathways in their cerebrum, as discussed earlier, are like uncharted territories, waiting to be formed and solidified.

During the first few weeks, a puppy's cognitive functions are primarily geared towards basic survival - feeding, staying warm, and bonding with their mother. But as the weeks roll on, their brain undergoes rapid changes. By the fourth week, they start recognizing familiar faces, both canine and human, and by the seventh week, they are capable of learning basic commands.

This period is also marked by 'critical periods of socialization.' It's a window where puppies are most receptive to new experiences, making it essential for owners to expose them to varied environments, sounds, and beings. The lessons learned during these moments have lasting impacts, shaping their personalities and behaviors in adulthood.

Adolescence to Adulthood: Refinement and Mastery

Lucy's teenage phase, much like human teenagers, was a blend of exuberance, rebellion, and discovery. Spanning from six months to about two years, adolescent dogs experience a surge in hormones, which can sometimes lead to challenging behaviors. However, cognitively, they are refining what they've learned as puppies, mastering commands, and understanding their place in the pack.

As they transition into adulthood, their cognitive abilities plateau in terms of growth but peak in terms of capability. An adult dog can understand and respond to a plethora of commands, recognize and remember various beings, and even exhibit problem-solving abilities. For instance, Lucy, when faced with a closed door, quickly learned to press down the handle with her paw, a testament to her evolved cognitive prowess.

Senior Years: The Grace of Wisdom

As the years start to weigh on them, dogs, like all beings, face a decline in cognitive functions. Beginning at around seven years, depending on the breed, senior dogs might start displaying signs of Canine Cognitive Dysfunction Syndrome (CCDS), akin to dementia in humans. They might appear confused, struggle with commands they once knew, or even display altered sleep patterns.

However, it's essential to recognize the grace that comes with this phase. Every gaze holds a depth of wisdom, every response, although slow, is measured and thoughtful. Their cognitive journey might be waning, but it's rich with memories, lessons, and experiences.

To aid our senior companions, mental stimulation becomes paramount. While it might not restore their youthful vigor, it helps in keeping their brain active and engaged. Simple games, puzzles, or even revisiting old tricks can make a world of difference.

In conclusion, the cognitive journey of dogs, from their boisterous puppy days to their contemplative senior moments, is a testament to nature's design. Each phase, with its highs and lows, paints a compelling story of growth, learning, and adaptation. As custodians of these incredible beings, recognizing and understanding these stages not only deepens our bond but also allows us to provide them with a life filled with understanding, patience, and love.

Breed-Specific Cognitive Traits and Tendencies

Have you ever watched a Border Collie herding sheep with astonishing precision or a Bloodhound tracking a scent with unwavering determination? These innate behaviors, deeply rooted in the fabric of their genetics, are more than just endearing antics. They're a reflection of the intricate tapestry of breed-specific cognitive traits and tendencies that have been honed over centuries.

When we talk about dog breeds, it's easy to get caught up in their physical attributes. However, beyond the fur colors and patterns, tail curls, and ear folds, lies a realm of cognitive diversity that's as vast as it is captivating.

The Border Collie: A Symphony of Focus and Intelligence

Take a moment and picture Martha, a sprightly Border Collie with a coat that glistens like a raven's wing under the midday sun. Martha isn't just playing fetch; she's strategizing. Every dart, twist, and turn is a calculated move. The Border Collie breed, originally developed for herding livestock, is known for its intense gaze, called "the eye." This laser-focused stare, combined with their unparalleled intelligence, makes them adept at understanding and manipulating the movements of a flock. The Collie's ability to anticipate and respond to shifts, often with mere gestures, is a testament to their breed-specific cognitive prowess.

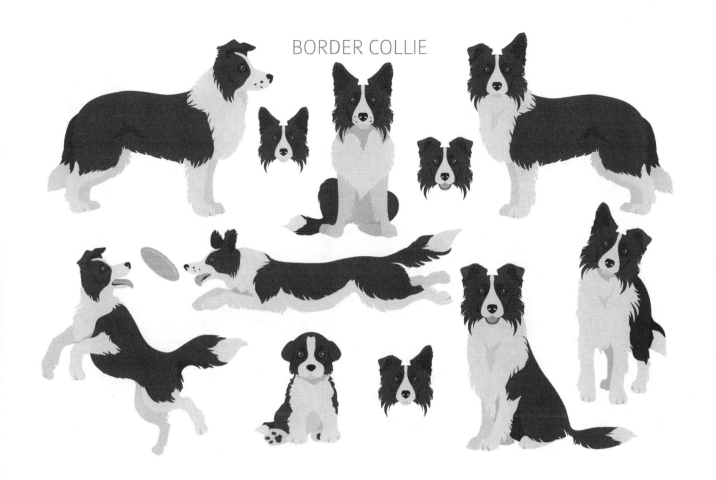

The Bloodhound: Scent Wizards with a Memory Map

Contrast Martha with Winston, a Bloodhound whose droopy eyes and sagging skin hide a world of sensory wonder. When Winston catches a whiff of a scent, his brain lights up like a bustling city at night. Bloodhounds have an extraordinary olfactory memory, allowing them to recall scents over incredible distances and after a considerable time. Their cognitive strength lies in their ability to discern and remember a myriad of smells, creating a mental map that guides them unerringly to their target.

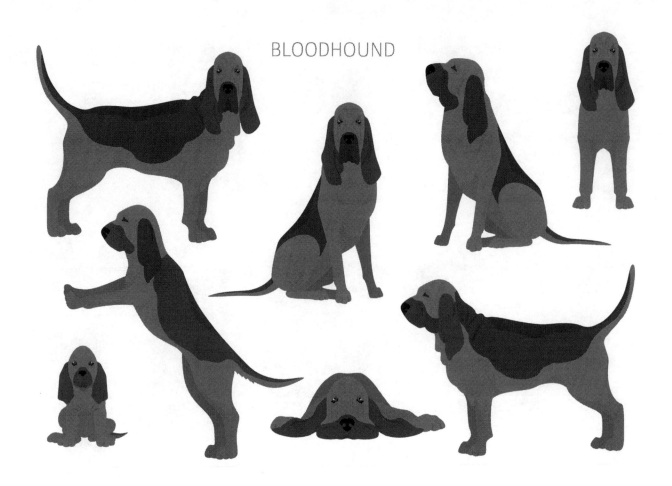

The Golden Retriever: Empathetic Companions

Then there's Sophie, a Golden Retriever whose gentle gaze seems to pierce right through to your soul. Golden Retrievers are not just about playfulness and loyalty; they exhibit a heightened sense of empathy. This trait, combined with their desire to please, has made them invaluable as therapy and service dogs. Their ability to read and respond to human emotions, to sense when someone is in distress and offer comfort, is a cognitive trait that's both heartwarming and invaluable.

The Dachshund: Fierce and Independent Thinkers

On the other side of the spectrum is Max, a feisty Dachshund with a bark that belies his size. Bred originally to hunt badgers and other tunneling animals, Dachshunds are independent thinkers. Their cognitive strengths lie in problem-solving, often displaying a tenacity that's both admirable and, occasionally, a tad exasperating for their owners.

Dachshund

This vast array of cognitive strengths, each tailored to a breed's historical role, paints a fascinating picture of canine diversity. But it's essential to remember that while breeds offer a blueprint, individual personalities can and do vary. Martha might prefer lounging to herding, Winston might choose cuddles over tracking, Sophie might have her mischievous moments, and Max might be a little cuddle bug.

In essence, understanding breed-specific cognitive traits offers a gateway into the world of our canine companions. It provides context, enriching our appreciation for their unique quirks and behaviors. But as with all things in nature, there's room for delightful unpredictability.

By diving deep into the cognitive tapestry of different breeds, we don't just become better trainers or caregivers; we become better companions. We learn to celebrate not just the shared heritage of a breed but the individual spirit of each dog.

In our journey with our four-legged friends, it's this blend of understanding and appreciation, science and love, that creates moments of magic. Moments where we don't just coexist but truly connect, transcending the boundaries of species, bound together by the threads of mutual respect and affection.

Our exploration of the canine brain and its myriad facets has taken us on a journey from the structured contours of its anatomy to the shifting sands of cognitive development and the colorful mosaic of breed-specific tendencies. It is a world where science meets soul, and understanding deepens our bond. As we move forward, let this knowledge be a lantern, illuminating the path of a rewarding, empathetic relationship with our four-legged companions. After all, in understanding them, we often discover more about ourselves.

Chapter 2: Recognizing the Signs

Imagine walking into a room bustling with energy, filled with delightful sights and curious scents; it's captivating, isn't it? Now picture that energy suddenly dissipating, replaced by a void of monotony. This contrast is what our canine companions face when they're mentally stimulated versus when they're not. Dogs, inherently curious and vivacious, thrive when their minds are engaged. However, when left unstimulated, this vibrancy can wane, leading to a cascade of symptoms and behaviors that often go unnoticed. Recognizing these signs is pivotal, not just for understanding our furry friends but for ensuring their holistic well-being. As we delve into this chapter, we will unravel the intricacies of an under-stimulated canine mind, the consequences of neglecting their mental needs, and the behavioral ramifications that arise.

Symptoms of a Mentally Under-stimulated Dog

The morning ritual of Maria and her young Dalmatian, Rico, changed dramatically over time. The joyous, energetic play she was used to witnessing every sunrise gradually shifted to Rico engaging in destructive behaviors—digging, chewing, and endless barking. These were not just signs of a mischievous dog; they were cries for help, signs of mental under-stimulation.

A dog's restlessness often screams volumes about its mental state. Dogs, especially those from high-energy breeds, may seem like they have boundless energy, pacing ceaselessly or playfully charging at household members. Without proper outlets or engagements, this energy may spiral into less desired outlets.

Destructive behaviors are common among dogs craving mental challenges. A favorite shoe chewed to bits, cushions torn apart, toys destroyed—not always signs of a "bad dog," but rather a pet trying to communicate its need for more engagement. It's essential for dog owners to decipher these acts not as mere mischief but as cries for help.

Along with destructiveness, increased vocalizations like excessive barking or prolonged whining might become prominent. While it's natural for dogs to bark or vocalize, a sudden spike in such behavior often indicates mental stagnation.

Some dogs manifest their boredom through repetitive actions—chasing after their tails, obsessively following a shadow, or running in familiar patterns. These actions might seem cute or quirky at first, but when they become a compulsion, they're clear indications of cognitive boredom.

Toys often serve as the bridge between dogs and their owners, a source of shared joy and play. When a dog starts losing interest in these toys, it can be heart-wrenching. It's not mere whimsy; it's a sign that the dog seeks deeper, more mentally engaging challenges.

One of the more challenging symptoms to handle is a dog's inability to settle down. Even after extensive physical play, they might appear restless, moving without purpose, or seeming perpetually on alert. Such behaviors, especially during quiet hours, can disrupt households and indicate a mind seeking more.

Lastly, a dog's diminished responsiveness or inattentiveness can be particularly concerning. When they no longer react to calls or commands, or when they seem distant, it's not defiance but a manifestation of their mental state, yearning for engagement.

These signs are the starting point. Recognizing them paves the way to deeper understanding and action. Like Maria did with Rico, one can transform the environment, introducing cognitive challenges, toys, and training that cater to a dog's mental needs. Understanding is the first step; acting on it deepens the bond, ensuring our beloved canines lead happy, balanced lives.

The Consequences of Neglecting Mental Exercise

When Sasha, a vibrant border collie, was adopted by the Thompsons, the entire neighborhood knew. Her boundless energy, her joyous leaps, and her warm snuggles made her an instant favorite. However, over the years, a shift was palpable. The radiant spirit started to dim, not from age but from a lack of mental engagement. The once social and loving Sasha began retreating, showcasing the severe consequences of neglecting mental exercise in dogs.

Many dog owners, especially new ones, can relate to the allure of physical activities—walks, playtimes, fetching games. They often form the core of the interaction, forging bonds between the pet and the owner. Yet, the pivotal role of mental exercises is less understood, often overlooked, to the detriment of the pet's well-being.

One of the foremost consequences of this oversight is the rise of destructive behaviors. Imagine having a vast reservoir of energy and no appropriate means to utilize it. For dogs like Sasha, this excess energy channels into activities like gnawing on furniture, digging up gardens, or tearing apart belongings. The damage is dual-fold: to the household items and, more crucially, to the dog's psyche, pushing it further into a cycle of frustration.

In addition, sleep disruptions become more frequent. Just as humans can't drift into a peaceful slumber with an overactive mind, dogs too grapple with sleep disturbances when mentally under-stimulated. Tossing, turning, whimpering in sleep, or even insomnia can become recurring issues, affecting their overall health and temper.

Furthermore, a dog deprived of mental stimulation often exhibits signs of weight gain. Without adequate cognitive challenges, they resort to constant snacking or overeating, stemming from boredom rather than hunger. This unhealthy weight gain can pave the way for an array of health issues, from joint problems to metabolic disorders.

The absence of mental engagement also has a profound effect on a dog's social skills. The once social and interactive dog might start displaying heightened aggression or become overly timid. These altered behavior patterns can affect their interactions not just with other pets but also with humans, including their beloved owners. The bond, once robust and full of trust, starts fraying, replaced by a chasm of misunderstanding and miscommunication.

Moreover, the lack of cognitive challenges often leads to anxiety-related disorders in dogs. The world, through their eyes, becomes a cacophony of unexplained noises, movements, and sensations. A car's honk, a visitor's footsteps, even the rustling of leaves can send them into a panic. This constant state of high-alert drains them, both mentally and physically, exacerbating their state of unease.

Another profound consequence surfaces in their training sessions. Commands once obeyed promptly become daunting tasks, or the dogs seem utterly disinterested. This decline in responsiveness can be baffling and frustrating for owners, not realizing it stems from the cognitive void their pets are experiencing.

Narrating Sasha's plight isn't to paint a bleak picture, but to underscore the paramount importance of mental exercise for our furry friends. Sasha, with time, understanding, and a reintroduction to cognitive challenges, managed to rekindle her old zest for life. The transformation was a testament to the magic that consistent mental stimulation can weave.

To safeguard our dogs from these consequences, awareness is the first step. Recognizing the signs, understanding the profound impact of mental exercises, and proactively engaging them in stimulating activities can spell the difference between a life of monotony and a life filled with vibrancy and joy. For our pets, who offer us unconditional love, it's the least we can do, ensuring their world is as enriched and fulfilling as they make ours.

Behavioral Issues Stemming from Lack of Mental Engagement

In a quaint corner of Brooklyn, Mrs. O'Reilly, a seasoned dog trainer, fondly remembers Rufus. Rufus wasn't her pet, but a canine student whose behavioral metamorphosis remains etched in her memory. A classic case of lack of mental engagement, Rufus was the epitome of behavioral issues that can sprout in dogs when their cerebral gears aren't regularly oiled.

To comprehend the entire spectrum of behavioral issues stemming from mental neglect, it's imperative to delve deeper, to understand the intricate tapestry of canine psyche and emotions. Dogs, much like humans, possess a range of emotions. When the mind remains stagnant, these emotions can convolute, manifesting in a myriad of behavioral concerns.

Take, for instance, compulsive behaviors. These are repeated, almost ritualistic actions that a dog might engage in, which don't serve any apparent purpose. Rufus, for example, developed an obsessive shadow-chasing habit. Hours on end, he'd be fixated, trying to catch the elusive shadow, a behavior that left him exhausted and the household perturbed. Such compulsions often mask the dog's craving for cognitive stimulation.

Similarly, there's a surge in attention-seeking behaviors. The once-independent dog might become excessively clingy, resorting to actions to constantly gain the owner's attention. Whether it's barking incessantly, nudging, or even engaging in actions they know are forbidden, the underlying cause is often an unsatiated mental appetite. For Rufus, it manifested in stealing items around the house, almost as if playing a desperate game of 'fetch'.

However, perhaps the most heart-wrenching consequence is depression. A lethargic demeanor, loss of interest in previously enjoyed activities, or even a decreased appetite point towards a dog grappling with melancholy. The vibrancy dims, replaced with an overwhelming sense of ennui. In a world where their stimuli are limited, the shadows of depression can cast long and enduring.

Increased aggression is another alarming behavioral change. The docile dog might now snap, growl, or even bite without apparent provocation. This heightened aggression isn't necessarily a testament to their character but a scream for mental enrichment. An idle mind, clouded with frustration and confusion, often finds solace in aggression, making it a defensive mechanism rather than an inherent trait.

Contrastingly, some dogs retreat into a shell, showcasing heightened timidity. Places, sounds, or even individuals they were previously comfortable with might now send them scurrying to a hiding spot. This heightened anxiety is a testament to their world becoming increasingly overwhelming, primarily because their cognitive faculties aren't being harnessed.

For many dogs, an absence of mental tasks leads to an incessant need for self-stimulation. Be it excessive licking, leading to sore spots, or even self-inflicted harm, the drive is to find an outlet, any outlet, to channel their pent-up cognitive energy. Rufus, at his lowest, started biting his tail, a behavior that ceased only when his mental needs were addressed.

Furthermore, an intriguing consequence of the lack of mental engagement is the erosion of learned behaviors. Commands and tricks once mastered seem forgotten. Not because the dog is unwilling but because the neural pathways, if not frequently treaded upon, can become obscure. The vibrancy of their training dulls, painting a picture of a dog that's regressed.

Mrs. O'Reilly's intervention for Rufus wasn't merely about curbing his shadow-chasing or tail-biting. It was about filling the void, the mental chasm that had manifested these behaviors. With puzzles, challenges, and tasks that engaged Rufus's mind, the transformation was palpable. The shadows no longer held allure, and his tail was safe. Rufus's journey from chaos to calm was a testament to the profound impact of mental engagement on behavior.

In the heart of every behavioral issue lies a plea, a silent beckoning for understanding and action. For our canine companions, the world is a mosaic of scents, sights, and sounds. When their ability to engage with this mosaic is curtailed, the world becomes a confusing labyrinth. Yet, with understanding, patience, and consistent mental engagement, the labyrinth can transform into an adventure, filled with joy, learning, and unbridled happiness.

Chapter 3: Laying the Foundations for Effective Play

Every relationship, be it among humans or between a person and their furry companion, is built upon foundational principles that deepen the bond over time. Among these, play stands out as a powerful tool to nurture and solidify this bond. But for play to truly blossom into a meaningful interaction, there are underpinnings we need to understand and integrate. The dance of positive reinforcement, the symphony of commands that go beyond the rudimentary, and the art of cultivating trust all converge to set the stage for transformative play.

The Psychology of Positive Reinforcement

Stepping into a room, you're met with the sound of gentle laughter, the sight of wagging tails, and the ambient aroma of freshly baked treats. At the heart of this delightful scene is a timeless principle: positive reinforcement. It's the invisible thread that weaves together the rich tapestry of canine training and is the foundation upon which meaningful connections with our four-legged companions are built.

Picture yourself as a child, handed a delectable piece of candy each time you finished your homework. You'd likely feel a rush of motivation to tackle your studies. Dogs, while distinct from humans in countless ways, share our propensity to respond favorably to rewards. In essence, positive reinforcement is the art and science of highlighting desired behaviors by introducing something pleasant or removing something unpleasant.

As trainers and passionate dog lovers, understanding the mechanics behind positive reinforcement is imperative. At its core, it's about capturing that golden moment when a dog does something right and instantly rewarding it. Whether it's a delightful treat, a heartwarming praise, or a game of fetch, the essence remains the same: reward the behavior you want to see more of.

When Charlie, a vibrant Golden Retriever, sits on command, he's not just obeying for the sake of obedience. In his canine cognition, the act of sitting has been associated with a joyous outcome, maybe a treat or a loving pat on the head. Over time, Charlie learns that this specific behavior yields pleasant results. The real magic lies in the speed and consistency with which the reward is delivered. A treat given five minutes post the desired act loses its efficacy. It's the immediacy of reinforcement that etches the behavior into a dog's memory.

But why does positive reinforcement wield such power? It's rooted in the basic psychology of learning. In the early 20th century, renowned psychologists like B.F. Skinner propounded the theory of operant conditioning, where behaviors followed by satisfying outcomes are more likely to be repeated. It's a simple, yet profound truth that transcends species.

Applying this to the canine world, when Luna, a sprightly Border Collie, fetches a ball and is instantly rewarded with a treat, her brain registers this pattern. The joy she derives from the treat makes her more inclined to repeat the fetching behavior. As these reinforced behaviors become ingrained, they shape a dog's overall demeanor, leading to a more harmonious relationship between dog and handler.

Yet, while the allure of positive reinforcement is undeniable, one must tread with care. Over-reliance on treats can lead to obesity or an overdependence on food as a motivator. The key lies in diversifying rewards. A harmonious melody of verbal praises, physical affection, and play can be just as effective, if not more, than treats alone.

Moreover, understanding a dog's unique preferences is pivotal. Just as we humans have our favored desserts and cherished songs, dogs too have their predilections. For some, a game of tug might be the zenith of joy, while for others, a gentle belly rub might just be the ticket.

But herein lies a question, a query that's echoed in many a dog lover's heart: Why not punishment? Why not correct a dog when it goes astray? The answer, rooted in a blend of psychology and empathy, is manifold. Negative reinforcements or punishments might momentarily suppress an undesirable behavior, but they fail to teach a dog the right way. Over time, punishments can foster fear, leading to a strained relationship between the canine and its human.

Conversely, positive reinforcement, bathed in the gentle glow of encouragement, not only shapes behavior but also nurtures trust. It's the bridge to a world where every 'sit', 'stay', and 'come' isn't just an act of obedience but a testament to a bond forged in love and respect.

In the realm of canine training, where countless methods vie for attention, positive reinforcement stands tall, not just as a technique but as a philosophy. It's an ode to understanding, a paean to patience, and above all, a tribute to the unwavering spirit of our furry companions.

In the subsequent sections, we'll delve deeper, moving beyond basic commands and exploring the bedrock of trust, the cornerstone of any meaningful play. But as we journey forth, let's carry with us the essence of positive reinforcement: a world where every command is a conversation, every reward a reaffirmation, and every moment a memory etched in the annals of companionship.

Essential Commands: Beyond the Basics

Picture the joy of teaching a young child their first word. Their wide-eyed wonder, the magic in their eyes as they utter that first syllable, the rapturous applause that follows. Now, imagine translating that same fervor into the world of canine training. In the beginning, there's 'sit', 'stay', and 'come'. But as with a growing child, the world of language for a dog is vast, untapped, and waiting to be explored.

Venturing beyond the basics is about elevating the conversation with our canine companions. It's about adding layers to their vocabulary, fine-tuning their skills, and deepening the bond that ties human to hound. This isn't just about complex tricks for show; it's about enriching their understanding of the world and the roles they play within it.

Take, for example, the command 'leave it'. At face value, it seems rather straightforward, a mere extension of 'no'. But dig a little deeper and you'll realize its multifaceted nature. 'Leave it' isn't just about getting your dog to avoid a piece of food on the floor; it's about self-control, patience, and understanding boundaries. In a situation where a curious puppy might find itself drawn to a potentially dangerous item, this command is a lifesaver.

Or consider 'heel', a command that often remains relegated to the shadows of dog training. Teaching a dog to 'heel' – to walk by your side without pulling on the leash – isn't merely about establishing control. It's about creating harmony during walks, about ensuring the journey is as enjoyable for them as it is for you. It's about transforming what might be a chaotic tug-of-war into a synchronized dance between two partners.

However, the intricacies of these commands aren't rooted in their complexity but in the nuances they bring to the table. It's one thing to teach a dog to 'fetch', but another entirely to teach them to fetch specific items. Imagine the magic of asking Bella, a vivacious Labrador, to fetch your slippers and watching her do so with precision. This isn't a mere game; it's a testament to her cognitive abilities, her understanding of objects, and her desire to engage in meaningful interactions.

Delving into these nuanced commands, though, requires patience and understanding. It's akin to teaching a child arithmetic after they've mastered counting. There will be stumbling blocks, moments of doubt, instances where it seems the message isn't getting through. But perseverance is key. Just as a child's eyes light up when they solve their first math problem, the joy in a dog's eyes when they master a complex command is palpable.

One might wonder, why venture into this deeper realm of training? Why not stop at the basics? The answer lies in the world of opportunities it unlocks for a dog. Advanced commands pave the way for mental stimulation, challenge their intellect, and offer them a richer, more diverse life. It's the difference between reading a child the same book every night and introducing them to a vast library of tales.

To embark on this journey, one must adopt a mindset of continuous learning. The world of canine training is ever-evolving, with new techniques and commands emerging regularly. But the bedrock remains the same: understanding, patience, and mutual respect.

Training is more than just rote learning; it's a dialogue. It's about listening as much as instructing. Each dog is unique, with their quirks, idiosyncrasies, and ways of perceiving the world. What works for Bruno, an exuberant Beagle, might not resonate with Max, a contemplative Greyhound. The key lies in customization, in tailoring the approach to suit the individual dog.

For instance, using toys as motivators might work wonders for a playful Pomeranian, while voice modulation might be the key to a sensitive Spaniel's heart. It's this fine-tuning, this dance of understanding and adaptation, that transforms training from a chore into an art.

Beyond the commands and techniques, though, lies a deeper truth. Venturing beyond basic commands is about more than just tricks and obedience. It's a love letter to our canine companions, a testament to the depths of our bond, a celebration of the journey we undertake together. It's about looking into their eyes and realizing that the conversation is boundless, that the horizon is vast, and that together, there's no command too complex, no challenge too daunting.

In essence, as we lay the foundations for effective play, it's vital to remember that our dogs aren't just pets; they're partners. Partners in play, partners in life, partners in the dance of existence. And as with any dance, it's not about the steps but the connection, the rhythm, and the joy of moving in harmony.

Building Trust: The Cornerstone of Interactive Play

There's a moment, fleeting yet profound, where everything falls into place. A shared glance, a playful nudge, or an affectionate lick – these are the hallmarks of a bond that's grounded in trust. If you've ever felt the heart-swelling emotion of your dog placing their paw gently in your hand, you'll understand the depth and breadth of this bond. Trust is not a commodity; it's a tapestry, intricately woven over time, stitched together by shared experiences and deep-rooted understanding.

The universe of dog training and interactive play orbits around this axis of trust. It's the invisible thread that ties commands to obedience, actions to rewards, and play to fulfillment. And while many elements of dog training might seem technical and methodical, building trust is pure art – raw, emotional, and profoundly intimate.

Imagine a scenario: You're at the park, a ball in hand, and your furry friend eagerly awaiting the toss. The air is thick with anticipation. As you throw the ball, your dog dashes after it with gleeful abandon. Now, many would see this as a mere game, a simple fetch. But peel back the layers, and there's a dance of trust at play. Your dog trusts that you'll throw the ball. They trust that this game has a purpose. And when they bring it back, dropping it at your feet, they trust that their efforts will be acknowledged and rewarded.

But how does one build this trust? Is it an inherent quality, or is it cultivated? The answer, much like the journey of trust itself, is multifaceted.

Fostering a Safe Environment

The first brushstrokes on the canvas of trust involve creating a safe, nurturing environment for your dog. This means a space where they feel secure, where the specters of fear and uncertainty are banished. Remember, trust cannot bloom in the shadow of intimidation or punishment. A dog that's constantly reprimanded, or worse, physically corrected, will harbor feelings of anxiety and apprehension. They'll second-guess their actions, always on edge, awaiting the next reproof. In such an environment, true interactive play remains a distant dream, for the play is built on mutual respect and understanding.

Consistency is Key

Imagine if the rules of a game kept changing. Today, a particular action earns applause; tomorrow, the same action draws ire. The result? Confusion, frustration, and eventual disinterest. The same principle applies to the realm of dog training. Consistency in rewards, in reactions, and in expectations lays the foundation for trust. When a dog understands the 'rules' and knows that they remain unchanging, they feel empowered, confident in their actions, and secure in the knowledge that their efforts will be recognized.

Listening with an Open Heart

Building trust isn't a one-way street. It's a dialogue, a continuous exchange of feelings, emotions, and intentions. And central to this dialogue is the art of listening. Not just with the ears, but with the heart. When your dog growls softly at a stranger, it's not mere aggression; it's communication. They might be feeling threatened, anxious, or merely protective.

Responding with immediate reproof might seem like the logical step, but it undermines trust. Instead, understanding the root of their behavior, acknowledging their feelings, and guiding them gently towards a more desirable reaction fosters trust.

Celebrate the Small Moments

Every leap in interactive play, every new command learned, every hurdle crossed, is a milestone in the journey of trust. Celebrating these moments, no matter how insignificant they might seem, strengthens the bond. It's not about extravagant rewards; often, a simple pat, a kind word, or a loving gaze is all it takes. These moments of shared joy, of mutual appreciation, are the stepping stones of trust.

Respecting Boundaries

Every individual, be it human or canine, has boundaries – invisible lines that define comfort zones. Building trust involves recognizing and respecting these boundaries. It's about understanding when to push and when to pull back, when to challenge and when to console. It's about realizing that every dog is unique, with their tapestry of experiences, feelings, and thresholds. Respecting these nuances, customizing the approach, and allowing them the space to grow at their own pace is intrinsic to trust.

In conclusion, trust is the lifeblood of interactive play. It's the bridge that connects human to canine, intent to action, and effort to reward. It's a journey, sometimes challenging, often rewarding, but always worth the effort. For at the end of this path lies a bond that's unbreakable, a connection that's profound, and a love that's eternal. So, as we venture into the realms of interactive play, let's remember that trust is our compass, our guiding light, leading us towards moments of shared joy, laughter, and fulfillment.

Chapter 4: Interactive Brain Games

Dogs are not just beings of physical prowess; they're thinkers, problem-solvers, and above all, curious explorers of their environment. This chapter delves deep into the realm of interactive brain games—a magical blend of play, challenge, and mental stimulation. While a simple game of fetch or a walk in the park can provide ample exercise, these activities sometimes just skim the surface of their intellectual capabilities. The beauty of interactive brain games lies in their dual nature: they're fun, yes, but they're also tools of enrichment. They spark that innate canine curiosity, asking our furry friends to think, decide, strategize, and above all, engage with the world around them in entirely new ways. Whether it's a twist on the classic hide and seek, an advanced fetch game that tests both their agility and intelligence, or puzzles that range from store-bought wonders to handcrafted enigmas, this chapter is a testament to the unlimited potential of play.

Hide and Seek Variations: Rediscovering an Age-old Game

Oh, the simple joys of childhood! Remember those afternoons where the world seemed to pause, and the only thing that mattered was finding the perfect hiding spot? Hide and seek wasn't just a game; it was an emotion, an adrenaline rush of being found, and the anticipation of finding. Now, picture this: what if we could recreate this pure, unbridled joy for our canine companions? The joy they would feel, searching for their beloved owner or favorite toy. Welcome to the world of Hide and Seek Variations for our furry friends.

The Classic Version with a Twist

The classic hide and seek game we all know and love translates surprisingly well into the dog world. Begin by getting your dog to sit and stay. Ensure that they're calm and focused. Then, go and find a hiding spot. The first few times, make it relatively easy — perhaps just the other side of a door or behind a piece of furniture. Once you're in place, call out your dog's name or give them a command to come and find you.

When your dog finally discovers you (and trust me, the joy in their eyes will be worth the wait), reward them with praises, pets, or a treat. As they get better at the game, challenge them with more intricate hiding spots.

But here's the twist: instead of always being the one to hide, let your dog be the one sometimes. Of course, they won't exactly tuck themselves behind the couch, but you can use their favorite toy to represent them. Hide the toy and ask your dog to find it. This little role reversal keeps the game fresh and adds an element of scent detection.

Hide and Seek in the Dark

Dogs have an exceptional sense of smell, which far surpasses their sight. To up the challenge and engage their olfactory senses more, try playing hide and seek in the dark. Ensure there are no obstacles they might bump into, and watch them rely even more on their noses. This version isn't just fun; it's also a brilliant exercise for their mental faculties.

The Multiple Hiders Game

Introduce more players into the game. It could be family members, friends, or even a mix of people and toys. With multiple hiders, your dog will need to decide whom to search for first. It's a playful lesson in decision-making and prioritization for them.

The Outdoor Challenge

Take the game outside. Parks or backyards offer countless hiding spots, introducing different terrains and smells. Hiding behind trees, bushes, or even gentle slopes can make the game more challenging. Just ensure the area is safe and enclosed so that your dog can't wander off.

Hide, Seek, and Command

This variant is for those looking to combine obedience training with fun. Once you're hidden and your dog is seeking, introduce commands. For example, when they're close, ask them to sit or lie down before they can claim their 'victory.' This variation ensures that your dog remains focused, obeys commands even in a high-stimulation environment, and understands that they need to work (or obey) for their reward.

Making it Rewarding

The joy of finding you or the toy is, in itself, a reward for most dogs. But to make the game even more enticing, consider introducing treats. However, don't just hand them a treat every time. Instead, occasionally hide the treat with you, making the final discovery even more rewarding. Alternatively, if they're searching for a toy, sometimes placing a treat with the toy can enhance their enthusiasm for the game.

Hide and Seek: More than Just a Game

Playing hide and seek with your dog isn't merely about entertainment. It's a multifaceted activity that offers numerous benefits. One of the primary advantages is mental stimulation. The game challenges your dog's cognitive functions, making them strategize, decide, and utilize their senses to their utmost capacity. Additionally, the game encourages physical exercise, getting them moving. Whether played indoors or outdoors, the act of seeking provides them with much-needed exercise. Furthermore, every moment you spend playing with your dog strengthens your bond. They come to understand that you're part of their pack, and shared activities like these only deepen and solidify this bond.

Hide and seek, in its various avatars, offers an unmatched blend of fun, learning, and bonding. It's a game that evolves. As your dog gets better, you find newer challenges. And as they grow older, the game can be modified to be gentler yet still engaging.

In a world that often runs at a frenetic pace, hide and seek gives you those moments of pause, of joy, and of connection. It takes you back to simpler times while allowing you to make precious new memories. So, the next time you find a moment, hide away from the world, and let the game of joyous discovery begin.

Advanced Fetch Games: Beyond the Basic Throw

Whoever said fetch was just about throwing a stick and having your dog sprint after it hasn't truly tapped into the full potential of this age-old game. There's an art, an elegance, and a thrill to fetch that's waiting to be uncovered. Let's unravel this tapestry of fun, shall we?

The World of Object Diversity: Size and Shape

Imagine the wind brushing against your face and the gleam in your dog's eyes when instead of the predictable tennis ball, you unveil a variety of playthings, each with its own essence. The weight, the feel, the way it sails or bounces—each toy carves a different path through the air, leading to unpredictable and exciting chases. Maybe it's the graceful arc of a Frisbee or the unpredictable bounce of an oddly shaped rubber toy. With each new object, you're not just changing the game's physicality; you're igniting curiosity and enthusiasm in your pet.

Throwing with Panache: The Craft and Care

Now, while we love the classic throw, let's not confine ourselves to it. There's magic in our wrists, waiting to be unleashed. Have you tried a soft lob, allowing the object to catch air and descend in an enticing slow motion? Or a swift skimming throw, barely touching the grass, urging your dog to sprint with all their might? Yet, remember to always consider your furry friend's safety and comfort. No throw, no matter how skillful, should put them in harm's way.

Turning Fetch into an Obstacle Adventure

You remember those action-packed chase sequences from movies, right? Why not recreate a snippet of that for your dog? Introduce them to a world where they get to weave around garden chairs, dash under low-hanging branches, or even jump over soft, safe barriers. Every obstacle becomes a question: "How do I get to my toy in the quickest way possible?" Watch as your dog employs wit and agility, turning the straightforward game of fetch into a delightful dance of dexterity.

When fetch goes beyond the basic throw, it's no longer just a game. It's a narrative, a ballet, a challenge and a bonding moment. The grass becomes a canvas, the toy a quest, and each throw a new chapter in the tale you and your dog weave together. Dive into this enriched experience and watch the simple joy it brings to you and your beloved pet.

The Adventure of Different Terrains

Walking along the same path every day has its comfort, but as they say, variety is the spice of life. Introducing your dog to different terrains for fetch isn't just about changing the landscape—it's about embarking on a novel adventure together. Feel the sand slip beneath your feet at the beach, hear the crunch of autumn leaves in a forest, or the hushed whispers of grass in a meadow. Each terrain introduces your dog to a new world of scents, textures, and challenges. The ball might skid faster on a frozen pond or get buried in tall grass, turning every fetch session into an unpredictable story of exploration.

The Dance of Commands

Imagine the harmony between a dance partner leading and the other following. That's the essence when we infuse commands into our game of fetch. It's not just about "fetch" or "come back." It's the pause before the throw, the anticipation in your dog's eyes when you softly say "wait," or the pride in their stride when they "find" a hidden toy on command. It's a choreography of words and actions, where you lead, and your dog gracefully follows, making each game a shared performance.

Mysteries of the Night-time Fetch

As dusk embraces the world and the stars begin to twinkle, a new fetch experience awaits. Night-time fetch is not just the same game under the cover of darkness. It's about the allure of the unknown, the thrill of chasing a glow-in-the-dark ball, and the amplified sounds of night. Every rustle of the leaves, every distant hoot of an owl, adds to the ambiance. Your dog relies more on their hearing and sense of smell, sharpening their skills and adding a touch of mystery to the chase.

Perfecting the Return and Release

The joy in fetch isn't just in the chase; it's also in the triumphant return, the prize proudly presented. But sometimes, the release doesn't go as smoothly as we'd hope. It becomes a playful tug-of-war or a game of chase. Yet, with gentle encouragement and consistent cues, this too can become a seamless part of the dance. Celebrate every successful release with praises or treats. Let your dog know they've done well, turning the return and release into a perfect finale to your shared adventure.

In the end, fetch is more than just a game. It's a series of shared stories, a bond strengthened with every throw and return, and memories crafted in every terrain, command, and twilight chase. Embrace these nuances, and watch as each fetch session becomes a chapter in the delightful journey with your canine companion.

Puzzle Toys: From Store-Bought to DIY

Imagine a rainy day, with the pitter-patter of drops creating a symphony on the roof. Your dog gazes outside, their tail slightly drooping, with the eagerness for outdoor play dampened by the weather. But what if the adventure didn't end there? What if, within the confines of your home, a world of exploration awaited your furry friend? This world is none other than the realm of puzzle toys.

A Glimpse into the World of Store-Bought Puzzles

The pet industry, with its pulsating heart and keen observations, has recognized the profound need for mental stimulation in our canine companions. It's no longer about a simple ball or a stuffed toy. Today, the shelves are adorned with intricacies meant to challenge, tease, and engage the intelligent minds of dogs.

Take, for instance, the tiered puzzles. Layers of spinning compartments hide treats, only to be unveiled when your dog learns to maneuver them in just the right sequence. It's a ballet of paws, snout, and sometimes, that ever-inquisitive tongue, working together in harmony to unlock the prize.

Or perhaps you've come across the toys that wobble and bob, dispensing treats at unpredictable intervals. The unpredictable nature of these toys means every push, nudge, or bite brings forth a unique result, ensuring your dog remains engaged for longer durations.

However, the market's magic doesn't lie solely in its variety but in its adaptability. There are puzzles tailored for the novice, guiding them gently into the world of problem-solving. And then there are the enigmas meant for the seasoned solver, those who've seen and solved it all, seeking a challenge that truly tests their mettle.

Crafting the Challenge: The DIY Route

Of course, store-bought puzzles offer convenience, but there's a certain charm and personal touch in crafting something with your own hands, infused with the love and care only an owner can provide. DIY puzzle toys aren't just about the end product but the journey of creation, where every step is a labor of love.

A simple muffin tin, often relegated to the far corners of your kitchen cabinet, can transform into a delightful puzzle. Place treats in some of its cups and cover each one with tennis balls. The challenge for your dog? Uncover and relish the treats, distinguishing the rewarded cups from the empty ones.

Another favorite involves a cardboard box filled with crumpled newspaper or paper balls, with treats hidden within. It becomes an excavation site for your dog, a hunt for buried treasure amidst the rustling paper.

However, DIY isn't merely about using what's on hand. It's about innovation, about watching your dog, understanding their quirks, their likes and dislikes, and then crafting a puzzle that speaks directly to their soul. It's an ever-evolving art, adapting and growing with each shared experience.

The Impact: Beyond Just Play

The allure of puzzle toys isn't just the immediate joy they provide but the lasting impact on a dog's mental and emotional well-being. Each twist, turn, or nudge at a puzzle is a step towards sharper cognitive functions, improved problem-solving skills, and heightened adaptability.

Furthermore, these toys play a pivotal role in alleviating anxiety. In the absence of mental stimulation, dogs often resort to undesirable behaviors—be it the incessant chewing of your favorite shoe or the mysterious disappearance of socks. Puzzles provide a constructive outlet for this pent-up energy, channeling it into a productive and rewarding activity.

Moreover, for the elderly canines, whose physical prowess might not be what it once was, these toys offer a gentle yet engaging way to remain active. It's heartwarming to watch an old dog, with graying fur and slow steps, light up with the same youthful excitement when faced with a new puzzle challenge.

As we wrap up this journey into the realm of puzzle toys, it's essential to remember that at the heart of every game, every challenge, is a bond—a bond that strengthens with every shared moment. Whether you're navigating the aisles of a pet store or rummaging through your home for DIY materials, the ultimate goal remains the same: creating memories, cherishing the joy, and celebrating the ever-evolving relationship with your canine companion.

In the end, every puzzle solved isn't just a testament to your dog's intelligence but a chapter in the story of togetherness, a tale where challenges are faced, mysteries unraveled, and every moment is a treasured memory.

Chapter 5: Scent Work and Nose Games

In the vast tapestry of our world, nature has endowed certain creatures with gifts so profound, they leave us in awe. Among them is the dog's extraordinary sense of smell, an ability that transcends mere biology and enters the realm of the magical. Beyond the basic sniff-and-seek games we often play with our furry companions, there exists a higher echelon of challenges that tap into this magic—advanced nose work. In this chapter, we'll journey into the intricate mazes of scent trails and the riveting world of competitions, where canines showcase their olfactory prowess. But beyond the challenges and accolades, this is a tale of deepening bonds and shared experiences between dogs and their humans. So, let's embark on this olfactory odyssey, discovering, marveling, and celebrating every step of the way.

The Science of Canine Olfaction

In the soft light of dusk, imagine your dog pausing on a routine walk, nose twitching as it catches an unseen aroma on the breeze. To us, it's merely a fleeting moment, but for your canine companion, it's a story told through scents, a novel written in invisible ink. It's in this profound world of olfaction where dogs truly reign supreme.

Their ability to decode the world through scent is a marvel of nature, something that has long fascinated scientists, trainers, and pet owners alike. But what lies beneath this incredible olfactory prowess? Dive with me into the captivating science of canine olfaction, and let's unravel the mysteries together.

An Evolutionary Perspective

The tapestry of time, spread across eons, offers intriguing tales of survival, adaptation, and evolution. When we trace the lineage of our beloved canine companions, we find their ancestors navigating a world painted in scents. These primeval terrains were teeming with olfactory stories that told of potential dangers, trails of prey, messages from fellow pack members, or the allure of a potential mate.

Early canids didn't have the luxury of supermarkets or secure homes. Every meal, every safety measure, was hard-earned. Their world was a game of shadows and sounds, and the winners were often those who could decipher the aromatic tales the wind carried. As evolution shaped these animals, their noses became their most potent tool - a compass, a radar, and a communication device, all in one. Over countless generations, the art of survival became deeply intertwined with the art of scent, refining and enhancing their olfactory prowess to the marvel we see today.

The Emotional Connection to Scents

Have you ever caught a whiff of a childhood perfume or the distinct aroma of a long-forgotten meal and been instantly transported back in time? Scents have a profound way of evoking memories and emotions. Now, magnify this experience manifold, and you might come close to understanding a dog's emotional journey with scents.

For our canine companions, every scent carries an emotion. Behind those bright eyes and wagging tails, lies an intricate neural network where olfaction and emotion dance in a delicate ballet. The amygdala, a part of their brain deeply rooted in emotions, processes each scent, attributing feelings, memories, and reactions to them.

Imagine your dog sauntering through the park, catching the scent of another dog. It's not just 'another dog was here.' It's a cascade of emotions and information - 'a young, possibly anxious female dog, perhaps in good health, played here just hours ago.' Each aroma is a chapter of a story, filled with characters, emotions, and narratives, playing out in the vast theater of their minds. Through understanding this profound emotional connection, we don't just see the world as our dogs do; we feel it, crafting a bond that's truly immeasurable.

Scent Memory: A Dog's Personal Diary

Step into a room with an old, familiar scent, and it can transport you back to another time, another place. Now, let's amplify that experience multiple-fold, and we can begin to grasp the wonders of a dog's scent memory. Every smell they encounter isn't just a fleeting moment; it's a bookmarked page in the vast, rich diary of their lives.

Imagine a library filled with countless books, each volume brimming with tales. For dogs, each scent is a story meticulously archived in this library. The aroma of the first snowfall, the earthy smell after a summer rain, or even the distinct odor of a place they once visited - all these get stored, revisited, and cherished. This remarkable ability ensures they remember friends, foes, routes, and even potentially dangerous zones or items. It's like flipping through a personal diary filled with memories, emotions, and lessons learned.

Decoding Pheromones: Silent Conversations

Just as we humans rely on words and body language for communication, dogs have an intricate language woven from pheromones, the chemical signals exuded by all animals. These silent, invisible messages float around, waiting for the right recipient. For a dog, every other canine they encounter is like an open book, pages drenched in pheromones, revealing tales of age, mood, health, and even intentions.

Imagine walking into a room and, without uttering a single word, understanding the emotional states and intents of everyone present. That's somewhat the magic dogs experience daily. Through pheromones, a simple meeting between two dogs in a park becomes a silent conversation, a nuanced exchange of information. It's a whispered dialect, profound and intimate, guiding interactions, friendships, and even potential confrontations.

Environmental Influence on Olfaction

The world, with its myriad aromas, isn't a static tapestry for our canine companions. Instead, it's a dynamic, ever-shifting mosaic where environmental factors play a crucial role in shaping the olfactory experience. Rain, for instance, can elevate certain scents, turning the world into a fresher, more vivid canvas. On the flip side, strong winds might disperse aromas, making them harder to pinpoint.

A dog's olfaction isn't just about the inherent power of their nose; it's a dance with the environment. Seasons play their part too. The sweet blossom of spring, the dry dust of summer, or the decaying leaves of fall – each offers a unique aromatic profile, a different chapter in the olfactory storybook they navigate. By understanding these environmental nuances, we get closer to seeing – or rather, smelling – the world as our furry friends do, in all its vibrant, fluctuating beauty.

Setting Up Scent Trails and Treasure Hunts

In a world teeming with tantalizing aromas, our furry companions are seasoned explorers, effortlessly navigating through a maze of scents that paint a much more vivid picture than we could ever see with our eyes. But why limit this extraordinary talent to their daily walks or backyard romps? Setting up scent trails and treasure hunts can harness this natural ability, transforming it into a delightful game for both pet and owner. This game is more than just a playful diversion; it's a bridge to a deeper connection and understanding between you and your four-legged friend.

Crafting The Perfect Scent Trail

Creating a scent trail for your dog is akin to writing a riveting novel with intriguing plot twists. It needs a blend of familiarity to make it inviting, with dashes of unpredictability to keep the curiosity alive.

Start by selecting a scent. It could be a familiar one, like a treat they adore, or something novel and enticing, perhaps a new essential oil (safe for dogs) or a unique type of food. Begin the trail at a recognizable starting point, like their favorite spot in the house or yard. Gently drag the scented object along a pathway, ensuring it's not too straightforward. Think of it as a winding river with occasional detours, not a straight highway. Hide the scent's source at the trail's end, ensuring it's a rewarding find.

While the idea is simple, the nuances can make all the difference. Consider the environment. Are there competing smells that might distract them? How can you ensure the scent remains robust throughout their hunt? Perhaps dabbing a bit of the scent on cotton balls placed intermittently can maintain intrigue.

The Lure of the Treasure Hunt

A treasure hunt amplifies the thrill. Instead of just one primary scent to follow, imagine multiple scents leading to various 'treasures' or rewards. The treasures could range from their favorite toys, treats, or even places with a unique scent to intrigue their olfactory senses.

When setting up a treasure hunt, think layers. The first layer is the broader area – perhaps your garden or a segment of your home. Within this, create mini zones, each with a different scent trail leading to a treasure. Make sure the rewards are worth the effort – a delicious treat or a toy they've been eyeing for days. And remember, the journey is as crucial as the destination. The hunt should be engaging, with a mix of challenges and easier trails, ensuring they're never disheartened.

Building on Instinct, Crafting Connection

By now, you might wonder, why go through all this effort? The beauty of these games lies not just in the mental and sensory stimulation they offer to dogs, but in the bond they help foster. Observing your pet as they decipher the scent trails, watching their determination and joy, is a window into their world. It's an opportunity to understand them better, to appreciate the intricacies of their olfactory talents, and to share in their achievements, however small they might seem.

Each trail you craft and every hunt you curate is a chapter in the evolving story of your relationship. It's a testament to the effort you're willing to invest to enrich their lives, and the joy you derive from their happiness.

Challenges and Safety Tips

While scent games are fun and fulfilling, safety should always be a priority. Ensure that the scents and treasures used are safe for canine consumption or interaction. Keep in mind the physical capabilities of your dog; not all terrains or hiding spots might be suitable for every breed or age.

Moreover, always observe your dog during these activities. If they seem frustrated or stressed, it might be a sign that the trail is too challenging or that something in the environment is bothering them.

A New Chapter in Play

In a world where technology often dictates play, with various gadgets and toys available for pets, there's something pure and enchanting about scent games. It's a return to the basics, to the very essence of what it means to be a dog. It taps into their primal instincts, celebrates their unique capabilities, and provides a platform for genuine, heartfelt connection between pet and owner.

By setting up scent trails and treasure hunts, you're not just offering a game; you're crafting experiences, creating memories, and building a bond that's anchored in understanding, respect, and mutual joy. As they say, sometimes the simplest pleasures in life are the most profound, and with these olfactory adventures, you and your dog are all set to embark on many such delightful journeys together.

Advanced Nose Work: Challenges and Competitions

In the dance of the wind, where scents swirl and intermingle, there lies an intricate tapestry of stories that only the keenest of noses can decipher. Our loyal canine companions, with their unparalleled olfactory prowess, are nature's best detectives, instinctively tuning into this mosaic of aromas. But as with any talent, to truly shine, it needs to be honed, challenged, and celebrated. This is where advanced nose work, brimming with complex challenges and the thrill of competitions, enters the picture. Dive with me into this captivating world and discover how you can elevate your dog's scenting abilities to new heights.

A Higher Calling: The Essence of Advanced Nose Work

While basic scent games offer delightful engagement, advanced nose work delves deeper, pushing boundaries and testing limits. It's not just about identifying a familiar aroma or finding a hidden treat; it's a symphony of complexity, where dogs are trained to differentiate between multiple scents, often in environments laden with distractions.

Imagine a bustling park, where the scent of blooming flowers, other animals, and even food stalls creates a myriad of olfactory notes. Within this, your dog must zero in on a specific scent, discerning it with precision amidst the cacophony. That's advanced nose work. It's the difference between reading a sentence and understanding a complex novel, absorbing not just the words but the nuances, subtext, and underlying themes.

Crafting Challenges: From Training to Execution

Elevating your dog's nose work skills requires careful planning and a systematic approach. Start with introducing them to a wider range of scents. Essential oils (always ensuring they're safe for dogs), spices, and even certain fruits can be incorporated. The idea is to train them to identify these scents in isolation and then amidst distractions.

You can set up scenarios where multiple scents are hidden in a controlled environment, like your home or garden. Over time, increase the complexity. Introduce background noises, perhaps play some music or invite a few friends over. The goal is to simulate real-world distractions, teaching your dog to focus solely on the target scent.

The Competitive Arena: Stepping into the Limelight

As your dog becomes proficient, consider entering the realm of nose work competitions. These events, held by various canine organizations, are platforms where dogs showcase their olfactory talents, competing against others in a range of challenging scenarios.

Competitions often involve multiple rounds, each with increasing difficulty. Dogs might be tasked with identifying specific scents in a large area, discerning between closely placed scents, or even tracking a moving scent source. The exhilaration of these events isn't just in the possibility of winning but in the journey – the preparation, the shared moments of training, and the sheer joy of watching your dog revel in their natural talent.

The Bond Beyond the Game

What makes advanced nose work so enchanting isn't just the complexity or the accolades. At its heart, it's a celebration of the bond between a dog and its owner. It's about understanding, trust, and mutual respect.

In the throes of training or amidst the high stakes of competition, there will be moments of uncertainty, perhaps even frustration. Maybe a scent trail proves too challenging, or distractions become overwhelming. In these moments, your role transcends that of a trainer or owner. You become their anchor, their pillar of support, offering encouragement, understanding, and unwavering faith in their abilities.

Your gentle words, the reassuring pat, or even a shared moment of rest can reignite their spirit, reminding them (and perhaps even you) that beyond the game, it's the shared journey that matters.

Navigating the Challenges: Ethical Considerations and Safety

Advanced nose work, while rewarding, comes with its set of challenges. It's essential to ensure that the scents used are always safe for your dog. Avoid using anything that might induce allergies or irritations. Also, during competitions or training sessions, ensure they have adequate water and rest breaks, especially in warmer climates.

Furthermore, always be mindful of their emotional well-being. Not every dog will relish the competitive environment, and that's okay. The aim should always be their happiness and well-being over any accolade or achievement.

A Symphony of Scents: The Closing Note

In the grand tapestry of life, every experience, challenge, and achievement adds a unique thread, enriching the overall design. Advanced nose work is one such thread in your dog's life, weaving moments of challenge, triumph, understanding, and deepening bonds.

As you embark on this olfactory journey, remember that it's not just about honing a skill or winning a competition. It's a voyage of discovery, a celebration of your dog's incredible talents, and most importantly, a testament to the love, trust, and understanding that you both share.

Embrace the challenges, relish the highs, learn from the lows, and let this adventure strengthen the unbreakable bond you share with your four-legged companion.

Chapter 6: Agility and Physical Brain Boosters

In the realm of canine well-being, agility stands as a testament to both mental and physical prowess. It's a symphony of swift decisions, nimble movements, and unwavering trust between dog and handler. But beyond the spectacle, agility is an adaptable world. From your backyard setups to the incorporation of cognitive twists and ensuring every breed finds its rhythm, agility celebrates diversity. Dive into this chapter to explore how to make agility a holistic experience, personalized for every pooch's unique flair.

Designing a Home Agility Course

A sun-kissed morning, the cheerful chirping of birds, and the ecstatic energy of your canine companion, bounding joyously through a series of jumps, tunnels, and weave poles in your backyard. The vision is more achievable than you might imagine, and the journey to that vision starts with designing a home agility course that taps into both your and your dog's creativity.

Agility is more than just a sport; it's a dance between a dog and its handler, an intricate ballet that melds precision with enthusiasm. But before you can waltz your way to perfection, the stage must be set, and that begins with understanding the elements of an agility course and how to adapt them for the home environment.

Start by envisioning the space. Whether it's your backyard, a spacious living room, or a garage, each location has its unique advantages. Backyards provide ample room, while indoor spaces shield from unpredictable weather. Decide based on the space you have and the equipment you plan to use. Safety is paramount, so ensure that the ground is even, free from sharp objects, and suitable for your dog to run on.

Next, consider the agility basics: jumps, tunnels, weave poles, and contact obstacles. These are the pillars of any agility course. For home-made jumps, sturdy broom handles placed on makeshift stands will do the trick. Remember, it's not about how high your dog can jump, but the technique and confidence with which they approach each obstacle.

AGILITY

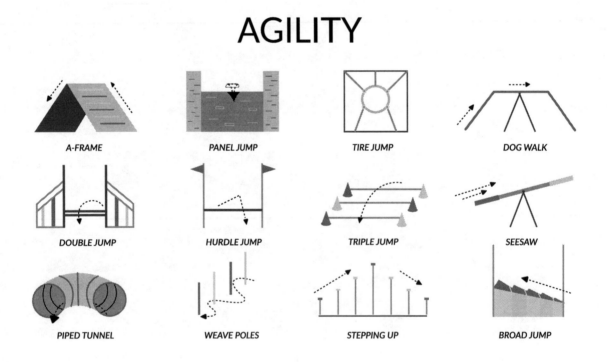

A-FRAME   PANEL JUMP   TIRE JUMP   DOG WALK

DOUBLE JUMP   HURDLE JUMP   TRIPLE JUMP   SEESAW

PIPED TUNNEL   WEAVE POLES   STEPPING UP   BROAD JUMP

Tunnels might sound tricky, but they don't have to be. An old barrel or children's play tunnel can be repurposed. If you're feeling particularly crafty, a series of wire hoops covered with a durable fabric creates a flexible tunnel that can be straightened or curved as desired.

Weave poles can be fashioned from PVC pipes planted in the ground in a straight line. The fun here is teaching your dog the graceful slalom-like movement through these poles. It's like watching a skier navigate a mountain course, each turn executed with precision and style.

Lastly, contact obstacles such as the A-frame or teeter-totter can be more challenging to replicate at home. However, with a bit of ingenuity, inclined planks with a grip surface (like a rubber mat) can simulate the experience. The goal is to teach your dog to navigate the obstacle while making contact with the "touch zones" at the start and end.

The magic of a home agility course lies not in replicating a professional setup but in crafting an environment where you and your dog can learn, bond, and grow together. It's about celebrating each small victory, be it the first successful weave through the poles or the jubilant dash through the tunnel.

When designing, let your imagination roam free, but always keep the comfort and safety of your dog in mind. Your living room lamp might look like a promising weave pole, but is it stable and safe? Your creativity is the limit, but caution is your compass.

In conclusion, the beauty of agility is the blend of physical exertion with mental stimulation. By creating a home agility course, you're not just setting up a playground; you're building an arena for memories, for laughter, and for the countless moments where you'll stand in awe of what you and your furry companion can achieve together. Just remember to have fun, be patient, and let the journey be as rewarding as the destination.

Incorporating Mental Challenges into Physical Play

The thrill of the chase, the spirited jumps, the agile dodges – these moments define physical play for our canine companions. But have you ever stopped mid-throw, that frisbee poised in hand, and wondered, "Is my dog mentally engaged, or just going through the motions?" Marrying the physical with the mental is not just a canine conundrum; it's an art that, once mastered, can transform playtime into a cerebral celebration.

When you think of mental challenges, perhaps the image that conjures is of a dog puzzle, treat hidden, with your pooch navigating the maze. But the genius of integrating the cerebral into the somatic lies in subtlety, weaving in challenges that your dog might not even recognize as such.

Picture this: It's a sunny day, and you're playing fetch in the park. Instead of the routine toss, try the "two-toy switch." Toss one toy and, as your dog returns with it, show them another, equally enticing toy. The mental challenge? They must drop the toy they have to chase the next. This not only sharpens their decision-making skills but also strengthens their impulse control.

Incorporate commands into play, turning them into a fun game rather than rote learning. For instance, during a game of tug-of-war, intermittently ask your dog to "drop it" or "leave it." When they obey, reward them by resuming the game. This not only reinforces training in a fun setting but also encourages them to stay mentally alert, waiting for the next command.

Then there's the "name game." Over time, teach your dog the names of their toys. Start with one, repeating its name every time they play with it. Once they associate the name with the toy, lay it among other toys, and ask them to fetch by name. Gradually expand their vocabulary. This game pushes their memory and association skills to the fore, making them recall names and associate them with specific objects.

Lastly, turn everyday walks into a treasure hunt of sorts. Scatter a few treats or toys along your walk route when your dog isn't looking. As you walk, encourage them to sniff and discover these hidden gems. It keeps them mentally engaged, as they're not only focusing on the walk but also anticipating the next delightful discovery.

Incorporating mental stimulation into physical play isn't about grand gestures. It's in the tiny moments, the split-second decisions, the anticipation, and the reward. The beauty of these combined challenges is that they not only ensure a tired dog (which as any dog owner will attest, is often a well-behaved dog) but also a mentally fulfilled one.

In wrapping up, it's essential to recognize that every game, every challenge, every new command learned during play is a testament to the deep bond between you and your dog. It's a dance of understanding, trust, and mutual respect. When physical play meets mental challenge, it results in a harmonious blend of body and mind, leaving both you and your dog enriched and ever more connected. Always remember, it's not just about the play; it's about the journey, the discoveries, and the shared joy along the way.

Adapting Agility for Different Dog Sizes and Breeds

Imagine a world where basketball players, no matter their height or reach, were asked to dunk on the same hoop height. It's not only impractical but borderline unfair. Similarly, when we delve into the world of canine agility, one size certainly does not fit all. Agility, at its heart, is a celebration of a dog's athletic prowess, their nimble footwork, their ability to weave, jump, and tunnel with grace and speed. Yet, to truly let every dog shine, the agility course must reflect the rich tapestry of dog sizes and breeds.

First and foremost, let's talk about our pint-sized champions - breeds like Chihuahuas, Pomeranians, or Dachshunds. Their shorter legs and compact bodies mean they can't and shouldn't be expected to jump the same heights as a Labrador or a Border Collie. For these little ones, the focus should be on lower jump bars, narrower weave poles, and smaller tunnels. And here's a pro-tip: with their closer-to-the-ground stature, they can often ace the low balance beams, showcasing their incredible center of gravity.

CHIHUAHUA SHORTHAIR

POMERANIAN

On the flip side, we have our larger breeds - think Great Danes, Mastiffs, or Saint Bernards. Their sheer size means they need wider jumps, spaced further apart. Weave poles should be set wider to accommodate their broader bodies. Tunnels, meanwhile, should be both wider and sturdier, ensuring they don't collapse under the weight. For these gentle giants, the agility course isn't just about physical adaptability, but also about teaching them spatial awareness, helping them understand their size and navigate the course accordingly.

Let's not forget our greyhounds and whippets, built for speed. These breeds, often termed sighthounds, have a natural proclivity for chasing, making them excellent agility candidates. For them, a longer track with spaced-out obstacles lets them build up speed and truly showcase their sprinting capabilities. Incorporate a few high jumps to let them exhibit their impressive vertical leap, but always ensure safe landings with cushioned mats.

Breeds like the Corgi, with their elongated bodies and shorter legs, present a unique challenge. It's vital to ensure that jumps aren't too high, putting undue strain on their spines. Instead, focus on weaving and tunneling activities, where their nimbleness truly shines.

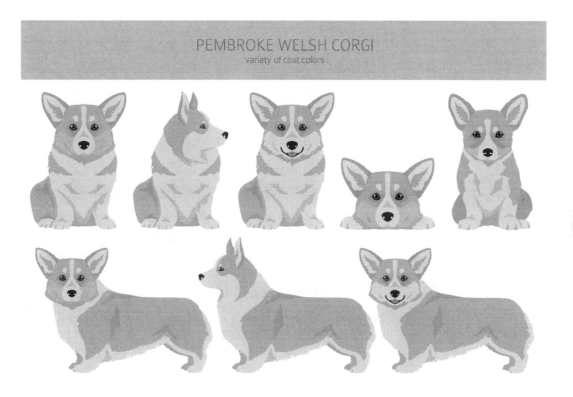

For breeds with herding instincts, such as the Border Collie or Australian Shepherd, capitalize on their natural desire to round things up. Integrate activities that tap into this instinct, like guiding a ball through a set path. Their laser focus and intelligence will come to the forefront, making it a joy to watch.

AUSTRALIAN SHEPHERD

It's also essential to consider the temperament of breeds. For instance, a Shih Tzu might be more tentative and require more coaxing through the course, while a Jack Russell Terrier might approach each obstacle with unparalleled enthusiasm. Recognize and respect these breed-specific quirks, adjusting the course and training style accordingly.

SHIH TZU
Chrysanthemum Dog

JACK RUSSEL TERRIER
smooth haired
variety of coat colors

In conclusion, agility is not about fitting every dog into a pre-set mold but allowing each breed and size its moment in the spotlight. Adapting agility courses for different breeds isn't a mere concession but a celebration of the incredible diversity in the canine kingdom. Whether they're weaving through poles, sprinting towards a finish line, or making that perfect jump, the key is to let each dog's unique strengths shine. In this dance of agility, every breed has its rhythm, its style, and its grace. Embrace it, celebrate it, and watch as every dog, from the tiniest Chihuahua to the majestic Great Dane, finds their moment of aerial poetry.

Chapter 7: Social Play and Group Challenges

In the intricate dance of life, social interactions form the very essence of existence, not just for humans, but for our beloved canine companions too. The world of dog play, filled with joyous barks and playful chases, is also layered with complex dynamics. From the exuberant leaps of confident dogs to the cautious steps of the shy ones, and even the misunderstood growls of aggression, understanding these nuances becomes pivotal. This chapter delves deep into the art of dog social play, exploring the beauty, challenges, and the means to enhance every playful encounter.

The Dynamics of Dog-to-Dog Play

The spark in their eyes, the wag of their tails, the playful stances; when dogs engage in spirited play with one another, it's not just a treat for the eyes but also an enriching experience for their social wellbeing. Delving into the world of canine interaction is akin to witnessing an intricate dance, laden with unspoken rules and subtleties that only they truly understand.

Just as humans have variances in social preferences and etiquettes, so too do our four-legged companions. Dogs, like us, communicate with a nuanced language, one that is filled with barks, tail movements, body postures, and facial expressions. Imagine entering a social gathering, where you're acutely aware of every slight shift in conversation tone or a fleeting glance exchanged. For dogs, every play session is this social gala, demanding alertness to myriad cues.

For many new dog parents or observers, distinguishing between friendly romps and potential scuffles can be a daunting task. In the world of dogs, play often involves behaviors that, to the untrained eye, might seem aggressive: chasing, tackling, and even some light mouthing. However, understanding the core dynamics can drastically alter our perceptions.

A prominent feature of healthy dog-to-dog play is role reversals. Today's chaser becomes tomorrow's chasee. This shifting dynamic ensures that no single dog dominates the play consistently, fostering a sense of fairness and balance. It's like children taking turns on a swing; each gets a fair shot at the exhilarating feeling of being airborne.

Then, there's the "play bow", a universal doggy signal saying, "Everything after this is in good fun!" This bow, where the dog stretches its front legs out and lowers its chest while keeping the hindquarters raised, is a clear invitation to frolic. It's the equivalent of a human waving and smiling, an unmistakable gesture of friendliness.

Yet, for all the rollicking and rumbling, dogs maintain a sense of boundaries. Pauses are interspersed in their interactions, like silent commas giving breath to a bustling sentence. These breaks allow dogs to process the ongoing play, ensuring things don't escalate into over-excitement or potential aggression. It's their own version of checking in, akin to a subtle "You okay?" during a vigorous dance.

Of course, not all dogs have the same play style. The terrier's enthusiastic and persistent chase differs from the gentle wrestle of two Great Danes. Each breed, and indeed each individual dog, brings its flavor to the play, much like people have varied dance styles, from the energetic jive to the graceful waltz.

Understanding these dynamics is crucial not just for the safety and happiness of the dogs involved but also for the peace of mind of their caregivers. When a dog parent can distinguish between a mouthy play gesture and a genuine nip or between an energetic chase and a predatory pursuit, they're better equipped to ensure positive experiences for their furry friends.

It's worth noting, though, that while most dogs cherish these interactions, not every dog wishes to be the life of the party. Some might prefer observing from the sidelines, some might engage in brief stints of play, and others might opt out entirely. And that's okay. Every dog, with its unique personality and preferences, deserves respect for its choices.

In the grand tapestry of dog-to-dog play, every twist, turn, and tangle tells a story. It's a story of connection, of communication, and of the pure, unadulterated joy of being in the moment. As caregivers, observers, and lifelong learners, our role is to understand, facilitate, and cherish these moments, ensuring our canine companions always have a dance partner ready for the next waltz.

Organizing Group Play Sessions and Challenges

Imagine a sunlit meadow, filled with the vibrant laughter of children, the hum of conversations, and the captivating smell of a nearby barbecue. Families gather, children play games, and there's an innate sense of community. Now, imagine a similar scenario, but with our four-legged friends at the heart of it, their joyous barks and playful antics stealing the show. Organizing group play sessions and challenges for dogs isn't merely about letting them run free. It's about cultivating a shared space where bonds form, dogs learn, and memories are made.

Creating such moments requires a blend of preparation, knowledge, and genuine love for the canine spirit. The playground – whether it's your backyard, a local park, or a dedicated dog play area – must be safe. Check for any potential hazards: sharp objects, harmful plants, or open water bodies where a playful chase might lead to an unintentional plunge. Remember, safety is paramount, much like you'd childproof a house for a toddler's first steps.

Once the environment gets a green signal, the next hurdle is group dynamics. As any party planner will tell you, a successful gathering isn't just about bringing individuals together; it's about ensuring they interact harmoniously. For dogs, this involves considering their temperament, play style, and past interactions. A sprightly young retriever might not be the best match for an older, calmer breed looking for a quiet corner. It's not about segregating, but rather about understanding preferences. If you've ever been to a social gathering where you felt out of place or overwhelmed, you'd appreciate the importance of such considerations.

Then comes the fun part – the challenges! Organizing games and challenges isn't merely about entertainment. It's an opportunity for dogs to learn, to test their skills, and to strengthen their bond with their caregivers. Consider a simple game of fetch, but with a twist. Two dogs, two balls, and a race to see who retrieves it first. Or, a maze set up with treats at various turns, challenging the dogs to use their sense of smell and wit. The aim isn't competition, but camaraderie, ensuring every dog, win or lose, enjoys the experience.

Of course, it's essential to have a watchful eye and an understanding heart throughout. Dogs, in their excitement, might forget boundaries. A challenge might become too challenging, leading to signs of stress or discomfort. Being observant ensures you can step in when needed, turning the situation around with a gentle word or a calming touch. It's about ensuring the laughter never fades, and the tails never stop wagging.

One might wonder, why go through all this effort? Is it not simpler to let dogs be, without the orchestrations and plans? To this, think back to the most memorable moments of your life. The surprise birthday parties, the meticulously planned trips, or the evenings where every detail was just perfect. We cherish these moments because someone made an effort, creating an environment where memories could be crafted.

For our canine friends, these group play sessions are their unforgettable evenings, their adventures, their stories to be retold in barks and howls. And at the heart of it all is you, the orchestrator of joy, the creator of memories, the one who understands that love, in its purest form, is about giving – time, effort, and moments of unbridled happiness.

In the world of wagging tails and playful barks, these group sessions are the symphonies where every bark is a note, every tail wag a beat, culminating in a harmonious melody of joy, learning, and togetherness. So, the next time you see a group of dogs playing, remember the silent maestro behind it all, ensuring the music never stops.

Addressing Social Play Issues: From Shyness to Aggression

In the soft glow of sunset, when the park is alive with the melodic cacophony of children's laughter and the distant chirps of returning birds, there's another symphony that often unfolds. The energetic dashes, playful growls, and delighted barks of dogs engaging in their evening frolic. Yet, within this tapestry of joy, there are subtle threads of tension, moments of hesitation, and the occasional outburst that seems out of place.

For anyone who's spent time observing dogs at play, it becomes apparent that social interactions among our four-legged friends aren't always straightforward. Just like us, dogs come with their own set of quirks, fears, and insecurities. Some are the life of the party, dashing from one playmate to another, while others hover on the sidelines, shy and uncertain. And then, there are those whose aggressive streaks cast a shadow on an otherwise sunny playdate.

Let's journey into the heart of these dynamics, understanding the roots of such behaviors and, more importantly, how we, as caregivers, can guide our furry companions toward healthier social interactions.

The Silent Observer: Addressing Shyness

If dogs could pen memoirs, the stories of the shy ones would be tales of longing, of watching from the shadows while their more confident counterparts took center stage. Shyness in dogs isn't merely a personality trait; it often stems from past traumas, lack of early socialization, or innate temperament.

To help a shy dog bloom, one must first offer the safety of understanding. Gentle reassurance, combined with gradual exposure to social situations, can work wonders. Picture this: A quiet corner of the park, early in the morning or late in the evening, when the crowds are sparse. Here, your shy companion can take baby steps into the world of social play, with you by their side, offering encouragement and comfort.

The Aggressive Playmate: Navigating the Storm

On the other end of the spectrum, we find dogs whose play often escalates into aggression. These aren't inherently "bad" dogs; instead, their aggressive tendencies might be a mask for underlying fears, territorial instincts, or past experiences that left scars.

Addressing aggression requires a blend of patience, observation, and timely intervention. By observing the triggers – be it a particular breed, a certain play style, or even a specific setting – one can predict and mitigate aggressive outbursts. In some cases, professional intervention, like that of a dog behaviorist, can provide invaluable insights and training techniques.

But beyond the strategies and interventions, there's a world of empathy waiting to be explored. Imagine for a moment being in their paws, overwhelmed by a world that sometimes feels too loud, too close, too much. Recognizing the emotions driving their behavior, be it fear, insecurity, or dominance, can pave the way for deeper understanding and effective solutions.

The world of dog social play is as complex and varied as our own social interactions. It's filled with joys, challenges, and opportunities for growth. As we guide our canine companions through this maze, it's essential to remember that every growl, every tail wag, and every playful dash is a window into their souls. And as we peer through, offering guidance, reassurance, and timely interventions, we aren't just addressing issues; we're building bridges, forging connections, and celebrating the beautiful, sometimes messy, dance of social play.

So, the next time you're in the park, watching the myriad stories unfold, take a moment to appreciate the silent narratives, the hidden challenges, and the triumphant moments. For in the world of dogs, every interaction is a chapter, every playdate a story, and every challenge an opportunity to grow, learn, and love more deeply.

Chapter 8: Advanced Training and Cognitive Challenges

In the intricate dance of dog training, we often lean on tradition. We're anchored to practices passed down through generations. But just as our world evolves, so do the ways we engage our canine companions. Advanced training techniques paired with burgeoning technological advancements promise not just to train, but to stimulate the canine intellect, painting a landscape where commands meet cognition, and playtime melds with innovation.

Clicker Training as a Brain Game

Amid the vast expanse of the canine training world, one tool stands out, shining with the brilliance of simplicity yet embedded with a profound impact. It's the clicker - an inconspicuous little device that, in the hands of a skillful trainer, can transform the way we communicate with our dogs. More than just a training tool, the clicker becomes a brain game, challenging our dogs mentally, honing their focus, and creating a symphony of understanding between two different species.

Picture this: a bright summer day, and there you are in your yard with your dog, armed with treats and a clicker. The journey begins with a simple 'click' followed by a treat. The underlying principle is as ancient as time - cause and effect. The click becomes a promise of a reward, a bridge over the gap between the desired behavior and the treat. It becomes the Rosetta Stone, translating human desires into a language the canine brain can decipher.

One might wonder, how can such a basic tool pose as a cognitive challenge? The answer lies in its application. In its essence, clicker training is akin to a dance, where timing, rhythm, and intuition play a pivotal role. The very moment your dog offers a desired behavior, the click sounds, capturing the behavior like a photographer captures a fleeting moment. This precision demands acute attention from the dog, transforming each training session into a mentally stimulating game.

Imagine trying to catch a delicate butterfly with a net; the timing has to be impeccable. For our dogs, the challenge is to figure out the exact moment, the precise behavior that triggers the sound of the clicker. It becomes an enticing puzzle, urging them to experiment, to try different actions, and to discern the pattern.

However, as with all tools, the magic isn't in the clicker itself but in the hands that wield it. The power of clicker training is amplified when combined with patience and creativity. Let's take the example of teaching a dog to spin. Initially, any slight turn might be captured with a click. As sessions progress, the criteria tightens; maybe now only a half spin earns a click. And then, finally, a full graceful spin. By raising the bar gradually, we're not just teaching a trick, but encouraging problem-solving, perseverance, and adaptability.

Moreover, the clicker opens up avenues to teach behaviors that might seem out of reach with traditional methods. From identifying objects by name to intricate dance routines, the sky becomes the limit. And every step of the way, it remains a game, a delightful puzzle that keeps the dog's neurons firing with excitement and anticipation.

However, amidst this praise, a word of caution. Like every powerful tool, the clicker must be used responsibly. One must avoid the temptation to overuse it, to ensure it doesn't lose its potency. Also, it's crucial to remember that the clicker is not a remote control but a communication device. The bond, the love, and the trust between the trainer and the dog remain at the heart of every successful training journey.

In conclusion, the humble clicker emerges not just as a tool, but as an orchestra conductor, bringing together the melodies of canine cognition, creativity, and communication. It's more than training; it's an art form, a brain game that deepens the bond we share with our four-legged companions, reminding us that the journey is as beautiful as the destination. As we embark on this adventure, let the click be our guide, echoing with the promise of understanding, growth, and joy.

Teaching Complex Tricks and Sequences

The world of canine training is vast, intricate, and filled with nuances. A single step beyond basic obedience or the initial fun trick lies the captivating realm of complex tricks and sequences. Venturing into this territory is like discovering the art behind the science, a dance that's choreographed meticulously and yet appears effortlessly fluid. Let's dive into this world where precision meets passion, and our bond with our canine companions reaches new heights of synchronization.

Imagine the elegance of a dog weaving through your legs as you walk, seamlessly transitioning into a backward circle around you, and finishing with a bow. This isn't just a spectacle for onlookers, but a testament to the countless hours of patient practice, the bond of trust, and the mental acumen of both the trainer and the dog.

Teaching complex tricks starts with understanding individual elements. Think of it like crafting a story. Each trick is a word, a sequence is a sentence, and a routine is a chapter. To convey a compelling tale, one must master the vocabulary first. For instance, before teaching our dog the intricate dance of weaving through our legs, we'd teach them to move around a single leg. This foundational trick, once established, can then be expanded upon, introducing variations and adding layers of complexity.

Now, sequences are where the true cognitive challenge for our canine partners lies. It's akin to a gymnast stringing together a set of maneuvers in a flawless routine. Each step must be executed perfectly and in the right order. For our dogs, this means recalling a series of actions and performing them in sequence, relying on cues from their human partner. An example could be the combination of a spin, followed by a leg weave, ending with a play dead. Such sequences not only provide physical exercise but stretch their mental capacities as they anticipate the next move, recall the correct action, and execute it with precision.

One of the most exhilarating aspects of teaching sequences is observing a dog's cognitive wheels turning. Their keen eyes trying to predict the next cue, their bodies tensed in excited anticipation, and that spark of understanding when they get it right is incomparable. This process amplifies their problem-solving skills, tests their memory, and enhances their focus.

It's essential, however, to approach the teaching of complex tricks and sequences with patience. Expecting perfection from the get-go can lead to frustration for both parties. Instead, relish in the incremental progress. Celebrate the small victories. The day your dog manages to weave through both legs, even if it's a bit clumsy, is a day for joy. Over time, with repetition and positive reinforcement, the clumsiness fades, replaced by elegance and grace.

Furthermore, respect your dog's individuality. Just as every human has unique talents and inclinations, so does each dog. While one might excel at agility-based tricks, another might have a penchant for tricks that require intricate paw movements. Recognize their strengths, work on their weaknesses, but most importantly, ensure that the process remains enjoyable for them.

In closing, teaching complex tricks and sequences to our dogs isn't just about the final performance. It's a journey that strengthens the bond, deepens mutual respect, and provides an avenue for both parties to learn, grow, and evolve. It's a dance of trust, understanding, and unspoken communication. So, as you embark on this fulfilling journey, remember to savor each moment, each misstep, each triumph, for in them lies the true essence of companionship.

Cognitive Toys and Tech for Dogs

In an age where technology continues to evolve at an unprecedented rate, it's no surprise that our four-legged friends have become beneficiaries of such progress. Today's market offers a medley of cognitive toys and tech innovations designed specifically for dogs, providing not just entertainment but an enriching mental workout.

Imagine a setting sun, casting its golden hue over the horizon. In the quiet embrace of the evening, there's a gentle hum of a device in your living room. It isn't the latest smartphone or a state-of-the-art speaker but a treat-dispensing toy that your dog is fervently interacting with. This scene captures the essence of the modern era where technology and canine enrichment seamlessly intertwine.

One of the groundbreaking introductions to the canine world has been interactive toys that adjust difficulty based on a dog's ability. These toys gauge a dog's interaction and challenge them just enough to keep them engaged without causing frustration. It's like handing them a puzzle, where pieces become slightly more complex as they get better at solving it. Such toys not only ensure prolonged engagement but also adapt to the learning curve of every individual dog, ensuring that no two experiences are identical.

Then there are toys that foster the natural hunting instinct of dogs. Imagine a toy that mimics the erratic movements of a squirrel or a rabbit, darting around unpredictably. For the dog, it's not just a chase, but a strategic game. They learn to anticipate, predict, and react, sharpening their cognitive reflexes in the process.

Incorporating technology into playtime has another unforeseen advantage: It allows pet parents to engage with their pets even when they're not physically present. Interactive cameras with treat dispensers enable owners to initiate a game of fetch or dispense treats from miles away. It's a heartwarming blend of technology and emotion, where distance no longer hinders the bond between a dog and its owner.

Furthermore, the tech industry's strides haven't just stopped at toys. We now have wearable tech for dogs! These devices monitor a dog's activity, provide insights into their behavior, and even offer suggestions for specific cognitive exercises based on their breed, age, and activity level. Imagine knowing when your dog is most active during the day or understanding the triggers that cause them stress. It's like having a window into their minds, offering an unprecedented level of understanding.

Now, with all this talk of technology, one might wonder, "Is it all beneficial?" As with every innovation, moderation and understanding are key. While these toys and gadgets offer immense benefits, they're tools to enhance, not replace, traditional play and training. They serve as supplements to the rich tapestry of activities and interactions that form the cornerstone of a dog's life.

It's also paramount to ensure safety. While these products undergo rigorous testing, every dog is unique. Monitoring initial interactions, understanding the toy's mechanism, and ensuring it aligns with a dog's temperament is crucial. The last thing anyone wants is for an enriching experience to turn stressful.

As we stand at the cusp of this exciting fusion of canine enrichment and technological advancement, it's heartening to think of the possibilities. Our beloved pets, who've been our companions for millennia, are getting a front-row seat to the marvels of the modern world. And as we introduce them to these wonders, we must remember to balance tech time with genuine, unfiltered moments of connection.

For in those moments — a shared look, a game of fetch in the golden glow of a setting sun, or a quiet cuddle on a rainy afternoon — lies the timeless essence of the bond between humans and dogs. Technology will evolve, toys will come and go, but this bond, pure and unchanging, is the true magic we must cherish.

Chapter 9: Crafting Tailored Play Routines

In the intricate dance of life with our canine companions, understanding their needs and tailoring routines to them is paramount. Chapter 9 unravels the art of crafting play routines, diving deep into assessing our dogs' progress, adapting play for different life stages, and most intriguingly, using play as a powerful tool to address behavioral challenges. This isn't just about fun and games; it's about harnessing play's transformative power, understanding its nuances, and utilizing it to foster a deeper bond with our four-legged friends.

Assessing and Documenting Your Dog's Progress

Engaging with your dog through games and training sessions isn't just about keeping them entertained; it's about fostering growth and understanding their evolving needs. When we set off on this journey with our furry companions, we're not just signing up for a joyride. Instead, we're committing to a dynamic partnership, built on trust and mutual respect. But how can we, as devoted dog owners, measure the progress made in this partnership? How can we truly understand if our teaching techniques and play routines are working?

Imagine you're learning a new language. After weeks of study, you're likely to assess your proficiency. You might test your ability to converse or understand, perhaps keep a diary of new words learned. Similarly, assessing and documenting our dog's progress helps transform our intuition into knowledge.

Begin by observing your dog during play sessions. Does Max now retrieve the ball more quickly than he did two weeks ago? Does Bella show less hesitation when presented with a new toy or a novel game? These are markers of progress. Their growth might not always be monumental leaps; sometimes, it's in the subtle ways they respond, the slightly improved reaction times, or the newfound enthusiasm they show for a particular game.

A practical way to track these subtle changes is by maintaining a 'Play Journal'. This isn't about rigorous bookkeeping, but a heartwarming diary of shared moments. Note down dates, the games you played, and any new reactions observed. Did your dog show a particular affinity to a new game? Write it down. Were they hesitant or scared about a new toy? Make a note.

For instance, one entry might read, "April 20: Introduced tug-of-war with a new rope toy. Bella was initially hesitant, perhaps the texture was unfamiliar. But after a gentle coaxing, she seemed to enjoy it. We played for fifteen minutes straight!" Over time, these journal entries create a rich tapestry of your dog's evolving play preferences and their journey of growth with you.

Documenting isn't just about recording reactions; it's also about capturing memories. Intermingle the functional aspects of your journal with photographs or little paw prints. A photo of Max's first successful fetch or Bella's jubilant sprint after mastering a command - these become cherished memories, making the journal not just a tool for assessment but a keepsake of beautiful moments.

Furthermore, this documented journey provides valuable insights. By revisiting older entries, patterns emerge. Perhaps during winters, your dog prefers indoor games, or maybe, after a specific training routine, their agility has notably improved. Recognizing these patterns allows for tailored play routines, ensuring your dog is always engaged and stimulated.

Lastly, remember that this process isn't about constant grading or seeking perfection. It's about understanding and bonding. Some days will be filled with rapid progress, while others might see minor setbacks. That's the beauty of growth; it's not linear. It's filled with peaks, troughs, and countless moments that make the journey memorable.

In the symphony of dog ownership, your Play Journal becomes a reflection of the shared rhythm between you and your canine companion. It's a testament to the effort invested, the joy derived, and the journey traversed together. By assessing and documenting, you're not just being an observant owner; you're honoring the relationship, marking the footsteps of a journey taken side by side.

Adapting Play for Puppies, Adults, and Senior Dogs

As the golden hues of autumn give way to the white frost of winter, trees shed their leaves in preparation for a new growth cycle. Nature understands that with change comes adaptation. And, much like nature, our dogs undergo various phases in their lifetime, each with its own set of needs and play requirements.

The Energetic Pup: Harnessing Boundless Energy

Consider the young puppy: a bundle of mischief, teething tendencies, and an insatiable curiosity for the world around. At this stage, play is more than just fun; it's a learning curve. Puppies are eager to explore and understand their surroundings. Soft chew toys can be a soothing relief for their teething gums. Engaging them in light tug-of-war games or gentle fetch helps them understand the basic concepts of retrieval and sharing.

For our tiny furballs, it's also essential to focus on games that stimulate their cognitive abilities. Hide and seek with toys or playing 'find the treat' engages their senses and instills problem-solving skills. But, remember, puppies are just starting to understand their body limits. Ensure playtime is short, fun, and allows for plenty of rest periods.

The Dynamic Adult: Perfecting the Balance

Transitioning to adulthood, our canine companions are at their peak – both physically and mentally. Here, play is not just about fun; it's about channeling their energy constructively. Retrieving games can become more intense, and hikes or long runs can be introduced. This is the prime time to teach them more complex tricks, be it dancing on their hind legs or fetching the morning newspaper.

However, each adult dog is unique. While some may have a natural affinity for water and would love nothing more than a splash in the pool, others might be content with a game of Frisbee. It's crucial to recognize their individual preferences and craft play routines accordingly. Tailoring play at this stage is about understanding their strengths and providing avenues to shine, all while ensuring they remain mentally stimulated and physically fit.

The Graceful Senior: Play with a Touch of Tenderness

As the sun sets, casting a warm, mellow glow, our dogs, now in their golden years, move at a gentler pace. But this slowing down doesn't equate to an end of play. Instead, it marks a shift in the nature of play. Senior dogs, while not as agile as their younger selves, are filled with a wisdom that only years can bring.

For our senior companions, play becomes less about physical intensity and more about companionship. Leisurely walks, gentle games of fetch, or simply sitting together in a park, enjoying the rustle of leaves, can be immensely satisfying. Cognitive toys, especially those that stimulate their senses without being physically demanding, can be introduced to keep their minds sharp.

One must be attentive to their comfort. Arthritis or joint issues are common in older dogs. Thus, games that exert undue pressure on their joints should be avoided. It's also essential to keep an eye out for signs of fatigue and ensure they get ample rest.

In the vast tapestry of a dog's life, each phase is a unique shade, vibrant and filled with its own set of joys and challenges. Adapting play routines to their life stage isn't just about keeping them entertained; it's a testament to our understanding and our commitment to ensuring their well-being, no matter the age. As guardians of their happiness, it is up to us to ensure that the play remains an integral part of their lives, evolving and adapting, just as they do. Through each bounce of the ball, each sprint, and each shared moment of stillness, we reaffirm our bond, ensuring it remains unbroken, from the playful days of puppyhood to the serene evenings of old age.

Addressing Specific Behavioral Challenges Through Play

In the grand symphony of a dog's life, there are high notes of joy and occasional discordant tones of behavioral challenges. Just as a skilled musician uses different techniques to strike the right chord, addressing canine behavioral issues often requires a thoughtful, tailored approach. Play, in this context, isn't just an activity; it's a transformative tool, a dance of understanding between dog and owner that can shift the dynamics of behavior.

The Tale of the Overly Energetic Mutt

Consider for a moment, Max, an exuberant Labrador with energy that seems boundless. His owners are often at their wits' end, trying to manage his hyperactive streak. Traditional training methods haven't yielded much success. Instead of resorting to strict discipline or confinement, they discover a different avenue: play.

Using a simple game of fetch, they channel Max's energy. But it isn't just any fetch game. They incorporate commands and pauses, making him wait, teaching patience, and redirecting his high spirits towards a constructive end. Over time, Max learns restraint without compromising his innate zest for life.

Mending the Bonds of Trust

Then there's Bella, a rescue with a traumatic past, uncertain of humans and wary of their touch. The scars on her soul manifest as anxiety and occasional aggression. While it's easy to label Bella as a 'problematic dog,' the truth is, she's a creature molded by her experiences, seeking safety.

Play becomes the bridge of trust for Bella. Starting with non-threatening toys, her owners engage her in gentle tug-of-war games. As she pulls on one end and they on the other, a silent conversation begins. Each tug is a question, each release an answer, and gradually, the walls Bella erected for her protection start to crumble. Through play, she learns that not all humans are threats, and it's okay to let her guard down.

The Case of the Territorial Terrier

And who could forget Rocky, the small terrier with a big personality, fiercely territorial of his space? Every time someone approaches his bed or toys, a growl emanates, a clear warning to back off. To address this, his owners introduce a playful strategy. They begin a game, nudging his toys with their feet lightly, then retreating, turning it into a playful dance of advance and retreat. Slowly, Rocky starts joining in, realizing it's all in good fun. The once-territorial terrier now playfully guards his toys, always ready for a game.

It's crucial to remember that every dog, like humans, comes with its own set of complexities. Labeling them as 'difficult' or 'untrainable' is a disservice to their potential. Play isn't just frivolous fun; it's a powerful medium, a language that both dog and owner can understand. It provides a safe space for dogs to express, learn, and transform.

Challenging behaviors, more often than not, stem from underlying issues. And while professional training and therapy play their part, never underestimate the magic woven by a simple game. As you toss that ball or engage in a playful wrestle, you're doing more than just entertaining your furry friend. You're teaching, guiding, and most importantly, bonding.

For every Max, Bella, and Rocky out there, there's a play solution waiting to be discovered, a key to unlock their best selves. As their trusted companions, it's up to us to find that key, to explore the power of play, and to remember that sometimes, the answers to our biggest challenges lie in the simplest joys.

As we journey through the process of tailoring play routines, one thing becomes clear: understanding and adapting to our dogs' individual needs is a cornerstone of a fulfilling relationship. Through careful assessment, age-appropriate play strategies, and addressing behavioral challenges with empathy and creativity, we pave the way for happier, healthier, and harmonious days with our beloved pets. Embrace the art of tailored play, for in it lies the key to unlocking a world of joy and mutual understanding.

Chapter 10: Overcoming Obstacles in Mental Stimulation

In the journey of canine companionship, every turn and trail presents its unique set of challenges. From the reluctant tail that hesitates at the sight of a new toy to the brave heart that has battled disabilities, every dog has its story. But, as with any saga, it's the manner in which we navigate these challenges that sets the tone. This chapter delves deep into understanding the intricacies of these obstacles and charts a path filled with safety, understanding, and joyous play.

Dealing with Reluctant or Fearful Dogs

Dogs, with their boundless enthusiasm and joy, often find happiness in the simplest of pleasures. However, behind the facade of excitement or nonchalance, some canines harbor emotions that stem from fear or apprehension. Introducing a reluctant dog to mental stimulation can resemble coaxing a child scared of the dark to brave an unlit hallway. The unknown, while intriguing, can be daunting.

A fearful dog might have several reasons for its apprehension. Traumatic past experiences, a history of neglect, or a simple lack of early socialization can make new toys or games seem more like threats than fun activities. Their world, from their vantage, is filled with potential perils. So, when a new toy or an unfamiliar game comes into view, the instinct to flee rather than play kicks in. The challenge then is not just about teaching a new game, but it's about reshaping their perception of the world around them.

To begin with, the bedrock upon which all other strategies stand is trust. A dog needs to trust its human. For a dog that is hesitant or fearful, understanding their emotion is the key. They're not avoiding play out of stubbornness; they're genuinely unsure or afraid. And for them, every small step towards trusting the new is monumental. Pushing them might set back progress, so it's vital to take it slow and celebrate the tiny victories. A slight nudge of a toy with their nose, an inquisitive sniff, or even a tentative paw touch could be their version of bravery.

A heartwarming story of overcoming fear comes from Luna, a rescue. The first time she encountered a treat-filled puzzle, she recoiled, barked, and retreated. To Luna, the toy was an enigma, possibly a threat. But with a sprinkling of treats around it and consistent, gentle encouragement, Luna went from being wary to being intrigued, and finally, engrossed.

For dogs like Luna, a phased introduction can work wonders. Before expecting them to play with a new toy, just let it lie around. Let the dog approach it in its own time, sniff it, and become accustomed to its presence. Over a few days, this passive familiarity can help reduce their reluctance to engage actively.

Using items or surroundings that the dog is already comfortable with can also aid in this transition. If a dog has a favorite toy or a particular spot they like, start there. The familiar smells and sensations can provide an anchor of comfort amidst the sea of unknowns.

Moreover, leading by example can be beneficial. Dogs are incredibly observant. By demonstrating how a toy works or how a game is played, we can pique their interest. Their innate curiosity, when paired with the assurance of their trusted human enjoying the activity, might just tip the scale towards participation.

In the end, while the path to mental stimulation might seem strewn with challenges for fearful or hesitant dogs, with patience and understanding, it becomes a journey of bonding, growth, and mutual trust. The goal is not just to get them to play a game but to broaden their horizons, one experience at a time. As they conquer each fear, they aren't just learning a new activity; they're rediscovering the world around them, and we get the privilege of being part of that beautiful journey.

Modifying Games for Dogs with Disabilities

Just as the sun paints the horizon with shades of gold and crimson, every dog, regardless of its physical limitations, has a spectrum of potential waiting to be unveiled. In the world of canines, disabilities don't define them; it's their heart, spirit, and will that truly characterize who they are. For these special dogs, mental stimulation is not just about play; it's an affirmation of their spirit, a testament to their ability to rise above their limitations.

Imagine the vibrant spirit of Oscar, a dachshund with paralyzed hind legs, radiating with joy as he cracks a puzzle toy using just his front paws and a dash of determination. Or picture the undying perseverance of Bella, a blind retriever, navigating her way through scent trails, relying solely on her heightened olfactory senses. These dogs might have disabilities, but their spirits, undeterred by challenges, seek out every drop of excitement and adventure from life.

When modifying games for dogs with disabilities, it's crucial to remember one golden rule: focus on their abilities, not their limitations. The world through their eyes is different, not lesser. It's a world where other senses might be heightened, where intuition plays a stronger role, and where the bond with their human becomes the guiding light.

Let's dive deeper into understanding how to adapt games for our special friends:

For dogs with mobility issues, like Oscar, the world might seem like a challenging maze. However, they often have a heightened sense of touch or a stronger upper body. Games that involve puzzle toys or toys that can be manipulated using front paws can be incredibly stimulating. Placing treats under cones or cups and encouraging them to lift or topple them can be both mentally and physically engaging. Furthermore, toys that produce sounds or have varied textures can make playtime more interactive and enjoyable for them.

Then there are our brave souls like Bella, who might lack sight but possess an exceptional sense of smell. For such dogs, games that rely on their olfactory skills can be especially rewarding. Creating scent trails using treats or specific essential oils can be an adventurous treasure hunt for them. Hide-and-seek, where they rely on their hearing and smell to locate their human or a treat, can be a delightful experience. Remember, it's about making them rely on their strengths.

Dogs with hearing impairments, on the other hand, often rely heavily on visual cues. Brightly colored toys, light-based games, or even simple activities like chasing the reflection from a mirror can be both fun and mentally stimulating for them. Their world is silent, but it's rich with visuals, textures, and scents.

Modifying games is not just about the activity itself; it's about the environment. Ensure that the play area is safe and familiar. Minimize obstructions and ensure that the surface is comfortable for them, especially for dogs with mobility issues.

As we adapt these games, it's essential to stay attuned to their responses. Every dog is an individual, and what works for one might not work for another. It's a dance of understanding, adapting, and evolving. The goal isn't to make them fit into the world of games designed for able-bodied dogs but to create a universe of play that revolves around them.

In conclusion, when we step into the world of dogs with disabilities, we aren't just trainers or pet parents; we are collaborators. Together, with love, patience, and creativity, we craft experiences that celebrate their spirit. We don't see their disabilities; we see their boundless potential, their resilience, and their undying will to embrace every shade of joy life has to offer.

Ensuring Safety and Well-being During Play

When the amber hues of evening descend, and the world takes on a gentler pace, there's a palpable shift in the air as dogs everywhere gear up for their favorite time: playtime. The infectious energy, the spirited barks, the animated tail wags - playtime is an unparalleled joy. But behind this symphony of happiness lies an essential undertone: the safety and well-being of our furry friends.

Picture this: Max, a boisterous golden retriever, is fully engrossed in a game of fetch. The ball soars, his eyes track it with unparalleled focus, and off he goes, racing with the wind. But in his fervor, he doesn't notice the sharp rock lying in his path. Moments like these underscore why ensuring safety during play is paramount.

First and foremost, the environment plays a pivotal role. Whether it's indoors or out in the open, the play area should be free from potential hazards. Uneven terrains, sharp objects, toxic plants, or even small items that can be ingested should be kept at bay. An environment audit, a simple walk-through of the play area, can make a world of difference. The ground should be stable, ensuring that dogs, in their exuberance, don't slip or hurt themselves.

Toys, the heart of many play activities, should be chosen with care. It's not just about the brightest or the noisiest toy; it's about what's safe. Toys that are too small can be a choking hazard, while those with detachable parts can be ingested, leading to severe complications. Durability is also key. A toy that can't withstand a dog's fervor can quickly become a hazard. And let's not forget about the materials. Non-toxic, pet-safe materials should be the standard.

As much as we focus on the physical, the emotional well-being of dogs during play is just as critical. Dogs, much like humans, have a spectrum of emotions. Overstimulation, frustration, or even fear can sneak into what's supposed to be a fun activity. It's essential to be attuned to their emotional cues. A dog that's showing signs of distress, be it through excessive barking, whimpering, or even aggression, might be signaling that something's amiss. Perhaps the game is too challenging, or maybe there's an element in the environment causing distress.

When introducing new games or challenges, it's advisable to take it slow. Much like the first brush strokes on a canvas, the initial moments set the tone. A gentle introduction, peppered with positive reinforcements like treats or verbal praise, can make the experience enjoyable. It's about building confidence, step by step.

Hydration and rest are the unsung heroes of play. In their zest, dogs can often forget to take breaks, leading to exhaustion or dehydration. Regular pauses, coupled with easy access to fresh water, ensure they remain at their energetic best.

Lastly, post-play checks can be invaluable. A quick once-over, checking for any signs of distress, injuries, or fatigue, ensures that any issues are nipped in the bud. This ritual not only ensures safety but also strengthens the bond, reinforcing the trust and love between the dog and the owner.

In the tapestry of dog ownership, play stands out as one of the most vibrant threads, a dance of joy, energy, and love. But it's a dance that should be free from the shadows of harm or distress. Ensuring safety and well-being during play isn't just a responsibility; it's a testament to the love and care that every dog deserves. As we revel in the joyous barks and the spirited games, let's pledge to make every playtime a safe, happy, and memorable experience.

As the curtain falls on this chapter, it's evident that the heart of canine mental stimulation lies not just in games and activities, but in the empathy and care with which we approach them. Overcoming obstacles is less about finding quick fixes and more about understanding, adapting, and ensuring safety. Because at the end of the day, our furry companions seek not just play but a bond that stands strong in the face of challenges, a bond that says, "No matter the hurdle, we'll leap it together."

Chapter 11: Beyond the Home: External Adventures and Challenges

Beyond the confines of home lies a world teeming with opportunities for canine exploration. It's an expansive stage set for adventure, bonding, and uncharted challenges. From the subtle transformations of an ordinary walk to the exhilarating rush of a new city's bustle, each environment promises unique experiences. This chapter delves deep into external adventures, ensuring your dog not only witnesses but thrives amidst these ever-changing backdrops.

Enriching Walks and Outdoor Adventures

The sun breaks through the morning mist, and as you lace up your sneakers, you can hear the unmistakable jingle of your dog's collar, filled with anticipation for the adventure that awaits. For many of us, taking our canine companions for a walk is a cherished daily ritual. However, these walks can be so much more than just a physical exercise. They can become a carnival of scents, sights, and sounds that mentally stimulate and deeply satisfy your dog's curious nature.

Imagine, for a moment, the world through your dog's eyes and nose. Every rustling leaf tells a story, every gust of wind carries tales from distant places, and every patch of grass is a chapter waiting to be read. While our human senses might categorize these as mundane, for our dogs, they're a symphony of information. So, how can we make these outdoor escapades even more enriching?

Firstly, let's delve into the idea of 'sniffari'. A term playfully coined by some dog enthusiasts, it emphasizes the importance of letting our dogs lead the way with their noses. Instead of hurrying them along, allow them to explore at their pace. This not only provides them with a wealth of sensory inputs but also gives them a sense of control and choice in their environment.

However, variety is the spice of life. Changing your walking route every so often introduces your dog to new terrains and smells. One day it might be a bustling urban setting, and the next could be a serene path by the lakeside. Each environment offers its unique set of challenges and stimuli. For instance, the urban setting might teach your dog to navigate crowds, while the lakeside path might introduce them to the fascinating world of waterfowls.

But walks can be educational too. Teaching commands like 'stop', 'go', and 'look' during these strolls can not only ensure safety but also add an element of structured learning. For instance, pausing at a busy intersection and asking your dog to 'look' both ways can become an enjoyable game that also instills the importance of being attentive.

Now, let's consider terrains. While flat parks and streets are standard, occasionally opting for hilly terrains, sandy beaches, or even rocky pathways can be both physically and mentally challenging for your dog. Figuring out how to navigate these different surfaces, deciding where to step or how to balance, all contribute to their cognitive development.

For those moments when you find a peaceful spot, maybe under a sprawling tree or by a gentle brook, you can introduce mini-games. Something as simple as a game of 'find the treat' where you hide a treat in your hand or around you and ask your dog to find it can be exhilarating for them.

In the grand tapestry of life, these walks are threads that weave moments of bonding, learning, and exploration. They remind us that sometimes, it's essential to take a break from the relentless march of time, to pause and immerse ourselves in the world around us. For in these pauses, in these moments of shared exploration, we discover not just the world, but also the depths of the bond we share with our canine companions.

In essence, each walk, each adventure, is an opportunity. An opportunity to strengthen the bond, to challenge the mind, and to discover the world together. As the sun sets and you both return, tired but content, know that today wasn't just another walk. It was a journey, an adventure, a story that you both will cherish forever.

Dog Sports and Competitions

The energy in the air is palpable, the anticipation tangible. Owners and their furry companions, each pair a unique testament to the bond of camaraderie, gather in the open field. Whispers of strategies, last-minute play sessions, and gentle words of encouragement echo around. It's not the Olympics, but for those involved, dog sports and competitions are just as thrilling.

Dog sports are a world of their own, where canine agility, intelligence, and obedience take center stage. It's not just about winning a trophy or getting a ribbon; it's about the journey, the bond, and the mutual respect formed between the handler and the dog.

Let's delve deep into this exciting realm.

Flyball, for instance, sees teams of dogs relay-racing against each other, jumping hurdles, and fetching tennis balls. The exuberance with which the dogs participate, their sheer joy at being able to sprint and leap, speaks volumes about the mental stimulation they receive from such activities. Here, it's not just the physical pace, but the need to understand commands quickly, to judge distances and heights, and to work as a team that challenges their intellect.

Then there's the obedience trial, a ballet of commands and responses. At the core of this sport is the bond between the handler and the dog. The duo moves through a series of tasks that range from following basic commands like 'sit' or 'stay' to more complex ones such as retrieving items or following scent trails. The challenge here isn't just in executing the task but understanding the nuances of each command, often conveyed through the slightest of gestures or changes in the handler's tone.

Agility trials, on the other hand, are a mesmerizing dance of speed, dexterity, and trust. Weaving through poles, darting through tunnels, and balancing on see-saws, dogs showcase their ability to navigate a series of obstacles with grace and speed. The mental challenge? Adapting to an ever-changing course and responding instantaneously to the handler's directions.

Yet, amidst the thrill of competition, one must remember that every dog is unique. While some may revel in the limelight, thriving on the energy of the crowd, others might find it overwhelming. It's essential to understand and respect these individual temperaments. The goal is enjoyment and mental stimulation, not undue stress.

Participation in these sports also fosters a sense of community. Handlers exchange tips, share stories, and often form bonds that go beyond the competition ground. It's a world where the love for dogs transcends all barriers, be it language, culture, or geography.

But perhaps the most beautiful aspect of dog sports is the celebration of the canine spirit. The tenacity of a terrier, the grace of a greyhound, the intelligence of a border collie - every breed, every dog brings something unique to the table. And as they jump, fetch, weave, or obey, they're not just showcasing their skills; they're expressing their joy, their love for life, and their unbreakable bond with their human.

In essence, dog sports and competitions are a testament to what dogs and humans can achieve together. It's a world where trust, respect, and love come alive, where every jump is a leap of faith, every command a promise of support. As you venture into this realm, remember, it's not about the trophies or the accolades; it's about the journey, the memories, and the tales of adventures shared with your furry companion.

So, the next time you hear of a local agility trial or a flyball competition, consider giving it a go. Even if you don't participate, just being there, amidst the energy, the camaraderie, and the sheer joy of the dogs, is an experience worth cherishing.

Travel and Exploration: Mental Stimulation on the Go

The sun casts its golden hue on the horizon, and the car engine purrs softly. You look over to the passenger seat, and there's your loyal companion, nose pressed against the window, ears perked up, absorbing the mosaic of sights and scents passing by. This isn't just any trip; it's an exploration, an adventure, a shared journey with your beloved dog.

Traveling with your canine companion offers an unparalleled avenue for mental stimulation. It's not merely about changing landscapes but introducing your pet to a plethora of experiences, from the rustling leaves in a distant forest to the rhythmic waves of an untouched beach.

Consider the seaside, for instance. For a dog, a beach isn't just sand and water. It's a treasure trove of sensations. The tactile feel of wet sand beneath their paws, the myriad scents of seaweed and salt, the sight of birds darting low over the waves, and the sound of the ocean's ebb and flow. Every element contributes to a rich tapestry of experiences, making each visit an exercise in sensory enrichment.

Similarly, think of a hike through a mountain trail. To us, it might be a test of endurance, a quest for a picturesque view. But for our furry friend, it's a cascade of discoveries. Every rock, every twig, every rustling in the underbrush presents an opportunity to investigate, to learn, and to grow.

However, travel isn't just about nature. Even urban environments can offer a wealth of experiences. The hustle and bustle of a city market, the echoing acoustics of a subway station, or the serene calm of a public park at dawn—all these contribute to your dog's understanding of the world, challenging them to adapt, adjust, and thrive amidst unfamiliarity.

That said, while the prospects of travel are enticing, they come with their fair share of responsibilities. It's not just about packing a bag and setting off into the sunset. When traveling with a dog, their comfort, safety, and well-being become paramount.

It's essential to ensure they are accustomed to the mode of transportation, be it a car, train, or even a plane. Familiarity can significantly reduce anxiety. A favored toy, a familiar blanket, or even the calming presence of their human can make all the difference in unfamiliar environments.

Adapting to new food and water sources, ensuring they are protected from local pests or diseases, and understanding the rules and etiquette of the place you're visiting—these are all crucial aspects of dog-friendly travel. A quick visit to the vet before any extended trip can provide peace of mind, ensuring that your pet is in the best of health and ready for adventure.

Yet, the true magic of traveling with your dog lies in the shared experiences. It's in those moments of serendipity—when you stumble upon a hidden meadow, when you watch the sunrise together in silent understanding, or when you simply sit by a campfire, lost in thoughts, with your dog resting its head on your lap.

In essence, travel and exploration with your dog go beyond mere recreation. They're a journey of the soul, an exploration of the bond that ties you two together. It's about growing together, learning from each other, and embracing the vast, beautiful world with open arms and eager hearts.

As you set off on your next journey, remember, it's not the destination but the journey itself that matters. The roads traveled, the memories made, the challenges overcome—they all weave into the tapestry of your shared life story, making each trip, each exploration, a chapter worth cherishing.

Our journey together through external challenges paints a vivid image: The world is not just a playground for our dogs but a classroom. Each adventure, whether a tranquil nature hike or a bustling urban escapade, molds them, enriches their senses, and strengthens our bond. As guardians of their experiences, we are given the unparalleled joy of watching them grow, adapt, and marvel at life's wonders. Together, let's pledge to explore, to challenge, and to cherish every moment beyond our homes.

Conclusion

As we journey through the winding paths of dog ownership, we often become hyper-focused on their physical needs: a balanced diet, routine vet check-ups, and daily walks. Yet, as the chapters of this book have illustrated, a dog's mental well-being is equally pivotal. It's not just about ensuring they are occupied; it's about kindling their inherent curiosity, challenging their intelligence, and nurturing their emotional depth. The process is transformative, with effects that ripple out further than one might initially realize. Our commitment to their cognitive enrichment doesn't just elevate their lives; it profoundly impacts ours, sculpting the world around us in subtle but significant ways.

The Lifelong Commitment to Your Dog's Mental Well-being

The sun sets, marking the end of another day. As you sit on your porch, you find your gaze drifting to your canine companion, curled up contentedly by your side. Their eyes meet yours, and in that brief exchange, there exists an entire universe of understanding and connection. It's a connection nurtured not just by love but by a profound commitment to their mental well-being. This bond, built over countless shared moments and mutual learning, speaks volumes about the lifelong journey you've undertaken together.

We've all heard of the saying, "A healthy mind in a healthy body." It's a universal truth, one that transcends species boundaries. As with humans, the physical health of a dog often takes precedence: routine vet visits, vaccinations, dietary choices, and daily exercises. But how often do we pause and contemplate the mental space our furry friends occupy? Understanding and catering to their mental well-being isn't a luxury—it's an absolute necessity.

First and foremost, the dedication to a dog's mental well-being is a testament to the depth of the bond shared. It's a silent vow we make, signaling our unwavering support throughout their life. Such a commitment comes from realizing that our dogs aren't just pets; they're family. We celebrate their joys, share their sorrows, and most importantly, remain attuned to the subtle shifts in their behavior and disposition. It's this heightened sensitivity that allows us to recognize when they need that extra bit of mental stimulation or a change in their daily routine.

Now, let's picture a scenario—a hypothetical yet all too common one. A young couple brings home a sprightly puppy, thrilled by its playful antics and boundless energy. The days are filled with laughter, games, and the joy of discovery. However, as the weeks turn to months, life's responsibilities start piling up. Work pressures intensify, personal commitments mount, and gradually, the once doted-upon pup finds itself receiving less and less attention. The toys gather dust, the games become infrequent, and the little furball, once the center of attention, feels neglected.

Such situations aren't uncommon, and more often than not, it's not out of malice but sheer oversight. Yet, the impact on the dog is profound. A lack of mental stimulation can lead to behavioral problems, anxiety, and even depression. Dogs, much like humans, thrive on engagement, challenges, and a sense of purpose.

So, what does a lifelong commitment to a dog's mental well-being entail? At its core, it's about consistency. Dogs revel in routines. It offers them a sense of security and predictability. Setting aside a dedicated 'play and learn' time daily, regardless of how busy life gets, goes a long way. It could be a simple game of fetch, a puzzle toy, or even just cuddling up with a book, with your dog resting beside you, basking in your presence. These moments, while seemingly insignificant, are the building blocks of a mentally healthy and happy dog.

Next comes adaptability. As dogs age, their needs change. The frisky games of their youth might give way to more relaxed activities. It's up to us to recognize these shifts and adjust accordingly. That might mean switching to softer toys, gentler games, or even just longer rest periods. The key lies in observation and responsiveness.

Education plays a pivotal role too. The world of canine mental health is vast and ever-evolving. New research, techniques, and methodologies emerge regularly. Making an effort to stay informed not only enhances our understanding but also equips us with the tools to offer our dogs the best possible care.

Finally, a commitment to a dog's mental health is about celebrating the small victories. Maybe it's the first time your dog manages to solve a puzzle toy or the way they've overcome a particular fear. Every milestone, no matter how minor, is a testament to the efforts you've both put in.

The journey might be long, filled with its fair share of ups and downs. But every step taken, every challenge faced, only serves to reinforce the bond you share. It's a relationship forged in love, trust, and a steadfast dedication to the mental well-being of your furry friend. And as the years roll by, as you both grow and evolve, you'll find that this commitment wasn't just beneficial for your dog, but deeply enriching for you as well.

In life, there are few joys as profound as seeing your dog happy, content, and mentally vibrant. And this, dear reader, is the heart of our message: That a lifelong commitment to your dog's mental well-being isn't just a responsibility—it's one of the most rewarding journeys you'll ever embark upon.

The Ripple Effects of a Mentally Stimulated Dog

The echo of a single pebble thrown into a pond can send ripples across the entire water body. Such is the nature of impact, where even the tiniest action can trigger a cascade of effects that resonate far and wide. When it comes to the mental well-being of our beloved dogs, the ripple effects are profoundly apparent, not just within the confines of our homes, but spilling into our communities, our relationships, and perhaps most importantly, within our own souls.

Imagine the sheer delight of a dog that has just cracked the code of a new puzzle toy, the gleam in their eyes, the wagging tail celebrating their accomplishment. That single moment of joy doesn't exist in isolation. It radiates outward, bringing a smile to our faces, a lightness to our day. We then carry that positivity forward, whether it's in a kind word to a neighbor, a more patient response to a colleague, or a gentler handling of our own internal struggles.

In the grand tapestry of life, every thread is interconnected. The mental vitality of our dogs doesn't merely translate to a happier pet; it fosters a more harmonious living environment. A mentally stimulated dog is less likely to exhibit destructive behaviors or lash out due to anxiety or pent-up energy. The curtains remain intact, the shoes unchewed, and the garden undug. But it goes deeper than material preservation.

When our dogs are mentally active and engaged, their behavior becomes more predictable, more in sync with the rhythms of our household. This predictability reduces stress, for both the pet and the owner. A reduced stress environment is conducive to better health, more meaningful interactions, and an overall enhanced quality of life. The benefits are tangible. There's less shouting, fewer accidents, and more moments of shared laughter and contentment.

Beyond the four walls of our home, the ripple effects continue to spread. A mentally alert dog is more social, better equipped to interact with other dogs in the park, and less likely to react aggressively or fearfully. This sociability paves the way for us, the owners, to form bonds with fellow dog lovers. Conversations spark, experiences are shared, and communities are forged on the foundation of mutual love for canines. The local park isn't just a patch of green anymore; it becomes a hub of camaraderie, a place where friendships blossom and support systems emerge.

Moreover, a mentally stimulated dog is a testament to responsible pet ownership. Neighbors, friends, and even strangers take note. It sets a precedent, a benchmark for others to emulate. "What's the secret?" they ask, intrigued by the calm demeanor and evident intelligence of the dog. And thus, knowledge spreads, awareness grows, and a collective shift towards better dog mental health practices gains momentum.

However, perhaps the most profound ripple effect is introspective in nature. The act of investing time and energy into the mental enrichment of our dogs becomes a mirror, reflecting back at us our values, our priorities, and our capacity for commitment and love. Every puzzle solved, every new trick learned, every fear overcome becomes a shared victory. These milestones, while centered around our pets, subtly shape our character, nurturing qualities like patience, empathy, and perseverance.

For in the end, isn't life itself a complex puzzle? Just as our dogs navigate the challenges we set for them, we too wade through the myriad puzzles life throws our way. And in those moments of uncertainty, of self-doubt, we find solace in the unwavering spirit of our canine companions. Their tenacity becomes our inspiration, their joy, our beacon of hope.

In conclusion, the journey towards ensuring the mental well-being of our dogs is not a solitary one. It's an endeavor that sends ripples across various facets of our existence, touching lives, molding relationships, and elevating communities. It serves as a powerful reminder of the interconnectedness of all things, of the boundless impact a single act of love and care can generate. The echo of that pebble, the ripple it creates, is a testament to the profound and far-reaching effects of a mentally stimulated dog. Embrace it, cherish it, and let those ripples spread, creating a symphony of positive change that resonates in every corner of our world.

Appendix A: Comprehensive Toy and Equipment Guide

Reviews, Recommendations, and Safety Tips

One thing is clear when you step into any pet store or peruse an online pet supply retailer: the world of dog toys and equipment is vast. Each brightly colored toy, every unique design, all seemingly scream, "Pick me! I'm the best for your pup!" But how can we, as dedicated dog owners, discern which toys stand above the rest, both in terms of quality and the mental stimulation they offer? Let's embark on this illuminative journey together, unraveling the complexities of the dog toy market while emphasizing the safety and happiness of our four-legged friends.

To begin, imagine the joy in a dog's eyes as they fixate on a new toy, the excitement evident in the wag of their tail, the eager anticipation in their stance. Now, contrast this with the crestfallen look they might have when that toy breaks too soon or, even worse, poses a danger to them. That's the spectrum we are looking at, and it's why making an informed choice is paramount.

Durability vs. Novelty

The first aspect to consider is the balance between durability and novelty. Toys designed to last might be made of robust materials like rubber or certain toughened plastics. However, the key is to ascertain if these materials are also safe for your dog. Some dogs, especially aggressive chewers, can tear apart even the most robust toys. In doing so, they risk ingesting harmful materials or choking hazards. On the other hand, a toy that's novel might engage your dog's senses in unique ways, such as toys that emit sounds, have interesting textures, or even give off scents. The challenge here is to ensure that these novel features don't compromise the toy's integrity or safety.

Interactive Toys – A Dual Purpose

Interactive toys are a blessing in disguise. Not only do they entertain, but they also challenge our dogs mentally. Think of toys that dispense treats when a certain puzzle is solved. These toys encourage problem-solving, but it's vital to ensure the mechanics of the toy are smooth, with no small parts that could be a potential choking hazard. Similarly, while it's exciting to find a toy that promises hours of engagement, it's crucial to see if any elements of the toy, especially the ones that make them interactive, are easily detachable or can be destroyed by a persistent pup.

Safety First

Beyond the build and mechanics of toys, it's important to consider other safety aspects. For instance, the paint or dye used on a toy: Is it non-toxic? Are there any sharp edges or points that could harm a dog if they chew or step on it? Always read labels and, if possible, go through customer reviews. Often, other dog owners' experiences can shed light on potential issues that might not be immediately apparent.

Personalized Choices

Every dog is an individual, with their own likes, dislikes, and play preferences. A toy that's perfect for a sprightly young Border Collie might not hold the same appeal for an elderly Basset Hound. Consider your dog's breed, age, health, and personality when making your choice. Some toys might be perfect for solo play, while others are more suited for interactive sessions with you.

Recommendations and Endorsements

It's tempting to rely solely on endorsements from famous personalities in the dog world. However, while these can be valuable, they shouldn't be the only factor in your decision-making. Combine expert opinions with real-world reviews from everyday dog owners. Sometimes, it's the mom-and-pop brands or lesser-known manufacturers that produce the gems in the market.

In conclusion, navigating the dog toy and equipment market can feel like a Herculean task. But armed with knowledge and a keen eye for detail, it becomes simpler. Always prioritize your dog's safety over any other aspect. Remember, the goal is to find toys that not only entertain but also enrich. As dog owners, our commitment is to ensure our beloved pets have a joyful, stimulating, and safe playtime. And with the right toys, that's a promise we can keep.

DIY Projects and Ideas for Home-Made Brain Games

The appeal of crafting something from scratch for a loved one is unparalleled. The time, effort, and love poured into creating a custom piece often render it more valuable than any store-bought item. Our dogs, with their uncanny ability to pick up on our emotions and intents, undoubtedly perceive this. Imagine watching your dog's eyes light up and tail wag in frenetic circles as they play with a toy you've made just for them. This section delves deep into the heart of DIY projects, focusing on creating mentally stimulating brain games for your canine companion right at home.

Crafting with Care

Starting on any DIY endeavor requires an acknowledgment of two critical aspects: safety and functionality. For our dog-centric creations, there's an added dimension: the fun and challenge it brings to the dog. Remember, the intention behind these homemade brain games is not just to entertain but to mentally stimulate and challenge.

The Essence of Materials

Selecting the right materials is the foundation of any DIY project. For our furry friends, it's important to pick non-toxic, dog-safe materials. Common household items like cardboard boxes, tennis balls, old t-shirts, and PVC pipes can be repurposed into fun and challenging games. Here's a guideline on some dog-safe materials:

- **Cardboard**: It's versatile, easily available, and safe for dogs. However, ensure that your dog doesn't end up eating chunks of it.
- **Tennis Balls**: Perfect for many DIY projects, but make sure they're clean and free of any small parts that could come loose.
- **Cotton Fabric**: Old t-shirts can be repurposed into tug toys or puzzle mats.
- **PVC pipes**: These can be transformed into treat-dispensing toys or intricate mazes. Just ensure all edges are smoothed out and there are no small detachable parts.

Ingenious DIY Brain Games

Treat-Dispensing Puzzles with PVC Pipes

For this, you'd need some PVC pipes, caps for the ends, and a drill. Attach the caps to either end of the pipe, ensuring a snug fit. Drill a few holes (big enough for your chosen treats to pass through) along the length of the pipe. Fill the pipe with treats and watch your dog roll and push it around, trying to get the treats out through the holes.

Tug-of-War Mat

Use an old t-shirt or two, cut them into strips, and weave these strips into a mat. Hide treats within the weaves, and your dog will use its nose and paws to tug and pull at the mat, trying to get to the treats.

Muffin Tin Puzzle

Take a muffin tin and some tennis balls. Place treats in some of the muffin slots and cover each slot with a tennis ball. Your dog will have to figure out which balls to remove to get the treats.

Box Tower

Stack cardboard boxes of varying sizes and hide treats inside some of them. Your dog will have to figure out which boxes to knock over or open to find the treats.

Scent Trails

Use your dog's favorite treats or some essential oils like lavender (ensure any scent used is safe for dogs). Create a trail or path by dropping the scent at various points, leading to a hidden treasure (a treat or favorite toy). This game engages your dog's primary sense – their sense of smell.

The Beauty of Customization

One of the most significant advantages of DIY projects is the ability to customize them to your dog's needs. If you have a senior dog, perhaps a game that doesn't require too much physical effort but still challenges them mentally would be apt. For puppies full of energy, games that require a combination of physical and mental efforts might be more suitable.

Cherishing the Moments

Above all, these DIY projects offer something even more precious than mental stimulation for your dog: they provide an opportunity to bond. The joy of creating something, combined with watching your dog enjoy and benefit from it, creates memories that will last a lifetime.

In Retrospect

Embarking on DIY projects for your dog, especially those that stimulate their brains, can be an incredibly rewarding journey. Not only are you ensuring their mental well-being, but you're also strengthening the bond you share with them. As with all things concerning our pets, safety is paramount. Always ensure that the materials used and the finished product are safe for your canine companion. Happy crafting!

Appendix B: Expert Insights and Further Learning

Interviews with Canine Cognitive Scientists and Trainers

Dogs are not just our faithful companions; they are individuals with their personalities, preferences, and cognitive processes. The world of canine cognition has seen remarkable strides in recent years, with researchers diving deeper into the minds of our beloved pets. To provide readers with a broader understanding, we had conversations with renowned canine cognitive scientists and seasoned trainers. These dialogues aim to blend academic understanding with practical insights from those who work hands-on with dogs every day.

A Conversation with Dr. Sophie Tremblay: Canine Cognitive Scientist

Understanding a dog's cognitive abilities is akin to unraveling a beautiful puzzle, begins Dr. Sophie Tremblay, who has dedicated two decades to studying the intricacies of canine cognition.

We often underestimate the mental capacity of our pets. Dogs don't just see the world as we do; they perceive it in a multitude of layers, particularly through scent, she elaborates.

Dr. Tremblay's research primarily focuses on memory and problem-solving in dogs. When asked about the most surprising aspect she's discovered, she excitedly shares, *Dogs have an incredible episodic memory. They can remember specific events, even if those events don't have direct consequences for them. This is something previously thought to be unique to humans.*

On the topic of mental stimulation, Dr. Tremblay cannot emphasize its importance enough. *Just as humans can experience cognitive decline, so can dogs. Regular mental exercises not only delay this decline but also enhance a dog's overall mental well-being,* she asserts.

Delving Deeper with Marco Stevenson: Veteran Dog Trainer

For someone who has worked with countless breeds and personalities, Marco Stevenson's passion for his profession remains undimmed. Marco, with his warm, infectious energy, has been training dogs for over thirty years.

The biggest mistake I often see is that owners approach training as a one-size-fits-all method, Marco states earnestly. He elaborates on how understanding a dog's individual cognitive style can drastically change training outcomes. *It's similar to teaching children; some might be visual learners, others kinesthetic. Dogs too have their styles, and recognizing that is half the battle won.*

Marco recalls a memorable experience with a particularly challenging Border Collie named Luna. *She was brilliant, too brilliant, in fact. Traditional training methods would bore her. It was only when I incorporated complex mental games that she began to respond.* This experience was an eye-opener for Marco about the importance of cognitive-specific training.

He advocates for an approach that combines both physical and mental training. *A mentally tired dog is a happy dog. They thrive when their brains are engaged. The spark I see in their eyes after a good problem-solving game is unmatched,* he says with a fond smile.

Insights from Clara Hughes: Specialist in Dog Behavior and Therapy

Clara Hughes has a unique perspective, having worked extensively with dogs that have faced trauma. Her work blends therapeutic interventions with cognitive exercises.

Dogs, like humans, can suffer from anxiety, PTSD, and other emotional disorders, Clara begins somberly. She recounts her experience with a Golden Retriever named Max, who had severe separation anxiety. *Traditional training did little for Max. It was only when we began incorporating mental exercises tailored to his needs that he started showing improvement.*

Clara emphasizes that understanding the dog's cognitive world is paramount. *Before we introduce any exercises, we need to understand their fears, triggers, and pleasures. Only then can we create a routine that genuinely helps them.*

Mental exercises, according to Clara, can be therapeutic. *They provide a sense of accomplishment to the dog. For dogs with trauma, this can be a step towards rebuilding their confidence,* she asserts.

Final Reflections

These conversations with experts from varied fields underscore a singular theme: the vast and intricate world within a dog's mind. Whether it's the academic rigor of Dr. Tremblay, the hands-on experiences of Marco Stevenson, or the therapeutic lens of Clara Hughes, one thing is clear—mentally stimulating our dogs is not just a luxury but a necessity.

Our journey with our dogs is enriched when we begin to see them not just as pets, but as beings with cognitive depths. By understanding their world, we not only ensure their well-being but also foster a bond that is deeply empathetic and understanding.

Recommended Reading, Courses, and Online Resources

In this age of digital information, dog owners have an abundance of resources at their fingertips. However, with the staggering amount of content available, sifting through and discerning valuable information can be a daunting task. In this section, we'll take you through a curated selection of books, courses, and online resources to deepen your understanding of canine cognition and enhance your bond with your beloved pet.

A Dive into Literature: Essential Books for Every Dog Owner

Journey through the Canine Mind by Dr. Alice Whitman is a transformative read that effortlessly combines rigorous science with the delightful storytelling of a dog lover. The book delves into fascinating realms of dog cognition, unraveling mysteries like the intricacies of their olfactory world and their understanding of human emotions. Dr. Whitman, with her knack for making complex ideas comprehensible, explores groundbreaking research findings, providing readers with actionable insights to better understand their canine companions.

Another gem, **Dogs and Us: Shared Worlds** by Neil Bronson, explores the co-evolutionary journey of humans and dogs. Bronson, an anthropologist and dog trainer, traces the roots of our symbiotic relationship with these loyal creatures. The narrative paints a vivid picture of ancient civilizations, where dogs played diverse roles - from hunters to guardians, and how these roles evolved over millennia. The essence of the book lies in its central message: the bond we share with our dogs is an age-old dance, one that has been refined and strengthened over countless generations.

For those looking for a blend of practical advice rooted in cognitive science, Lucy Henson's **Think Dog: Cognitive Training Techniques** is a treasure trove. Lucy, a seasoned trainer, harnesses her deep understanding of dog psychology to present a range of games and exercises. More than a manual, this book encourages owners to see the world from their dog's perspective, fostering mutual respect and understanding.

Beyond Books: Engaging Courses to Elevate Your Knowledge

The *Canine Cognition Center* offers an eight-week course titled **Understanding the Dog's Brain: From Puppies to Seniors**. Led by a team of experts, participants embark on an enlightening journey that begins with understanding a puppy's rapidly developing brain, travels through the cognitive peaks of adulthood, and delves into the changes occurring during a dog's golden years. Hands-on activities, interactive sessions, and real-life case studies make this course a favorite among dog enthusiasts.

For those keen on deepening their practical skills, **The Dog Whisperer Academy** presents *The Art and Science of Training. This course, spread over ten weeks, is a deep dive into effective training techniques underpinned by cognitive science. Participants have the unique opportunity to work closely with seasoned trainers, benefiting from their years of experience. The course is peppered with guest lectures from renowned canine cognitive scientists, ensuring a holistic learning experience.

Digital Domains: Trustworthy Online Resources

The digital age offers a plethora of platforms dedicated to canine enthusiasts. Among the most respected is **CanineMinds.org**, a hub for all things related to dog cognition. Regular webinars, featuring interviews with experts in the field, insightful articles, and a community forum make this platform an invaluable resource. Their monthly newsletter, packed with the latest research findings and practical tips, has become a favorite read for many.

For those seeking a blend of entertainment and education, the podcast **Dog Talk** hosted by veteran trainer and behaviorist Sam Green is a delightful listen. Each episode sees Sam conversing with guests from diverse backgrounds, be it researchers, trainers, or everyday dog owners. These candid conversations, interspersed with Sam's personal anecdotes, offer listeners a panoramic view of the vibrant world of dogs.

Lastly, **The Canine Club** is an online community that has rapidly gained popularity. This platform brings together dog lovers from across the globe, offering them a space to share experiences, seek advice, and collectively celebrate the joys of dog ownership. Their resource section, meticulously curated, features articles, videos, and tools, ensuring that members have access to reliable information.

The world of canine cognition, vast and vibrant, beckons every dog owner. As we journey through it, equipped with knowledge and a sense of wonder, we discover not just the marvels of our dog's mind but also deepen the bond we share with them. In this curated guide, each resource, be it a book, course, or online platform, serves as a beacon, guiding us through the mesmerizing landscape of our shared life with dogs.

Whether you're a seasoned dog owner or have recently welcomed a furry friend into your life, there's always more to learn and explore. Dive into these resources with an open heart and mind. You'll emerge not just with enriched knowledge but also with a renewed appreciation for the intricate dance of understanding, respect, and love that defines our relationship with our canine companions.

Appendix C: Case Studies

Real-life Stories of Transformation Through Mental Play

Lucy and Whiskey: A Second Chance at Connection

Whiskey, a golden retriever with a coat as rich as his namesake, was a rescue dog. Found wandering the streets, it took months of patience and care for Lucy, his new owner, to break through his shell. Whiskey was distant, exhibiting signs of past trauma. But when Lucy discovered the power of mental play, it became the bridge that brought Whiskey out of his shell.

Lucy first introduced Whiskey to puzzle toys. Initially skeptical, Whiskey's curiosity got the better of him, and soon he was enthusiastically solving puzzles to earn his treats. The spark in his eyes returned, and with every solved puzzle, Lucy watched as layers of his past trauma peeled away.

It wasn't just the toys, though. Lucy incorporated short but mentally engaging training sessions. Whiskey learned to associate commands with fun activities, and this process reshaped their daily interactions. Over time, trust blossomed between them, and Whiskey transformed from a dog haunted by his past to one living joyfully in the present.

Jasper's Journey with Ethan: Unlocking Hidden Potentials

Ethan's border collie, Jasper, was always high-energy – a trait characteristic of his breed. But when Ethan had to relocate for work, leaving behind their sprawling farm for a small city apartment, Jasper's energy became a challenge.

It was during a walk in the park that Ethan had an epiphany. Watching children play a memory game, Ethan felt inspired. He started teaching Jasper names for his toys, turning retrieval into a memory game. "Find Teddy," he'd say, and Jasper would search for the specific toy among a sea of others.

This seemingly simple game had profound effects. Jasper's restlessness subsided as he began to engage in focused, purposeful play. His bond with Ethan deepened as they spent hours challenging each other – Jasper's keen intelligence against Ethan's creativity in developing more intricate games.

Rebecca and Daisy: From Fear to Confidence

Daisy, a timid beagle mix, was adopted by Rebecca from a shelter. The little dog was fearful of everything – loud noises, new people, even her reflection. Rebecca, determined to help Daisy overcome her fears, turned to mental play.

Rebecca's strategy was genius in its simplicity. She started with mirrors. Placing treats near them, she encouraged Daisy to approach. With time and positive reinforcement, Daisy began to associate her reflection with treats and affection, dispelling her earlier fears.

Harnessing the power of scent games, Rebecca taught Daisy to track specific scents, turning their home into a maze of delightful discoveries. Daisy's natural tracking instincts took over, and her confidence grew with every successful find.

The culmination of their journey was a trip to the beach. What once would have been an overwhelming experience for Daisy became an adventure, with Rebecca setting up scent trails in the sand and Daisy eagerly tracking them, tail wagging, ears perked up in excitement.

Milo's Metamorphosis with Alex: From Chaos to Calm

Alex's terrier, Milo, was the embodiment of chaos. Leave him alone for a few minutes, and he'd turn the living room into a tornado's aftermath. Frustrated but not disheartened, Alex decided to try a different approach - mental play.

Alex started with 'hide and seek.' This game transformed their afternoons. Milo learned to stay in one spot while Alex hid, and then the hunt began. The excitement of the chase, coupled with the reward of finding Alex, turned their earlier destructive routine into a constructive, bonding activity.

But the true magic happened when Alex introduced Milo to 'imitation games'. Alex would perform an action, like spinning around or jumping, and Milo would imitate. This not only engaged Milo mentally but also allowed Alex to introduce commands gradually. The wild terrier learned control, responding to cues, and waiting for instructions, turning their once chaotic relationship into a harmonious dance.

Stories like these illuminate the transformative power of mental play. It's not just about keeping our dogs occupied; it's about unlocking a deeper layer of understanding, trust, and connection. Through mental play, barriers are broken, bonds are forged, and we come to see our canine companions not just as pets but as intelligent beings with vast emotional landscapes.

Every dog, like every human, carries a story. But with patience, love, and the right tools – like the power of mental play – we can rewrite these stories, turning tales of trauma, fear, or chaos into sagas of connection, understanding, and mutual growth. And in this journey of transformation, we, the humans, are changed just as profoundly as our beloved dogs.

Addressing and Overcoming Specific Behavioral Issues Through Mental Play

The Tale of Bella: Conquering Separation Anxiety

Bella, a lovely Dalmatian, always had a flair for drama. Every time her owner, Clara, stepped out, Bella's world seemed to crumble. Neighbors reported heartbreaking howls, and Clara would often return to scratched doors and torn curtains. Separation anxiety was the invisible chain binding Bella.

Determined to help, Clara delved into the world of mental play. She began with 'farewell puzzles,' special toys only given when Clara was about to leave. Bella's focus shifted from Clara's absence to solving these intriguing puzzles. Over time, the association of Clara's departures with positive play drastically reduced Bella's anxiety.

Max's Story: Overcoming Obsessive Behaviors

Max, a sprightly Corgi, had a fixation with his tail. Hours were spent in pursuit, often leaving him dizzy and distressed. Sarah, his owner, recognized this obsessive behavior and decided to intervene.

Drawing from the world of mental stimulation, Sarah introduced 'distraction games.' Whenever Max began his tail chase, Sarah would engage him in a voice-command game. "Where's your ball, Max?" she'd ask. Max, ever eager to play, soon began diverting his attention from his tail to his toys. This redirection of focus helped Max overcome his obsessive tail-chasing, giving him a healthier outlet for his energy.

Juno's Journey: Curbing Excessive Barking

Juno, a vivacious Beagle, had a voice and wasn't afraid to use it. While barking is natural, Juno's seemed excessive. Neighborhood walks became orchestras of barks, and postmen dreaded their delivery rounds.

Emma, Juno's owner, knew this wasn't mere mischief. Juno was communicating, and Emma needed to understand. Emma introduced 'quiet games,' a series of mental exercises promoting calmness. One such game was 'Whisper,' where Juno had to respond to commands given in a whisper. This game not only mentally stimulated Juno but also taught her the value of quietness.

Riley and the Battle with Aggression

Riley, a robust Rottweiler, had always been protective. But of late, his protective nature was morphing into aggression. Children playing nearby or other dogs approaching the house were met with a fierce growl and bared teeth.

Martin, Riley's owner, realized this aggression stemmed from Riley's heightened sense of responsibility to protect his territory. To combat this, Martin turned to 'role-playing games.' Enacting various scenarios, Martin taught Riley to differentiate between actual threats and harmless encounters. For example, by role-playing with a friend acting as a 'stranger,' Martin would use commands to teach Riley when to be alert and when to relax. These sessions provided Riley with mental stimulation while reinforcing appropriate behaviors.

Nina's Nights: The Struggle with Night-time Restlessness

Nina, a nocturnal Husky mix, turned nights into adventures. While the family slept, Nina prowled, often knocking things over or tearing up cushions. Clearly, her energy wasn't syncing with the family's routine.

Madison, determined to find a solution, tapped into the realm of mental play. Introducing 'sunset puzzles' – intricate toys designed to challenge Nina, Madison ensured Nina expended her energy solving them. By the time she was done, a sense of accomplishment and fatigue set in, aligning her sleep cycle closer to the family's.

Concluding Insights

Behavioral issues in dogs aren't just challenges; they're opportunities. They're windows into understanding our canine companions better, deciphering their unique ways of perceiving the world. By harnessing the power of mental play, we can address these behaviors, not through reprimands, but with understanding and guidance.

Mental play provides a platform for communication, allowing us to connect with our dogs, teaching them, and learning from them in return. Behavioral issues, when addressed through mental play, cease to be problems and become catalysts for deeper bonds.

In the symphony of life, every dog has its unique rhythm. Sometimes, they might go offbeat, but with the right tools, patience, and understanding, we can help them find their melody again. Through these tales of transformation, we realize the profound impacts of mental play – it's not merely a diversion but a path to harmony, understanding, and mutual growth.

Made in the USA
Monee, IL
17 December 2023

49751700R00077